T0332670

# Guidelines for Achieving Project Management Success

# Guidelines for Achieving Project Management Success

Gary L. Richardson and
Deborah Sater Carstens

CRC Press
Taylor & Francis Group
Boca Raton London New York

CRC Press is an imprint of the
Taylor & Francis Group, an **informa** business

Various quotes, figures, and tables are taken from our previous books, Project Management Tools and Techniques, 2nd edition (9780367201371) Project Management Theory and Practice, 3rd edition (9780815360711). Reproduced with permission from Taylor and Francis Group LLC (Books) US through PLSclear.

First edition published 2022
by CRC Press
6000 Broken Sound Parkway NW, Suite 300, Boca Raton, FL 33487-2742

and by CRC Press
2 Park Square, Milton Park, Abingdon, Oxon, OX14 4RN

© 2022 Gary L. Richardson and Deborah Sater Carstens
CRC Press is an imprint of Taylor & Francis Group, LLC

*Library of Congress Cataloging-in-Publication Data*

Names: Richardson, Gary L., author. | Carstens, Deborah Sater, author.
Title: Guidelines for achieving project management success / Gary L. Richardson and Deborah Sater Carstens.
Description: Boca Raton, FL : CRC Press, [2022] | Includes bibliographical references and index.
Identifiers: LCCN 2021040036 (print) | LCCN 2021040037 (ebook) | ISBN 9781032112350 (hardback) | ISBN 9781032112367 (paperback) | ISBN 9781003218982 (ebook)
Subjects: LCSH: Project management. | Management. | Business planning.
Classification: LCC HD69.P75 R5224 2022 (print) | LCC HD69.P75 (ebook) | DDC 658.4/04--dc23
LC record available at https://lccn.loc.gov/2021040036
LC ebook record available at https://lccn.loc.gov/2021040037

ISBN: 978-1-032-11235-0 (hbk)
ISBN: 978-1-032-11236-7 (pbk)
ISBN: 978-1-003-21898-2 (ebk)

DOI: 10.1201/9781003218982

Typeset in Times
by Deanta Global Publishing Services, Chennai, India

To Shawn and my growing family tree.

Gary Richardson

I dedicate this book to my loving children, Ryan and Lindsay. My parents, Stan and Norma, my brothers, Craig and Richard, and my sister-in-law, Pam, have always been encouraging.

Deborah Sater Carstens

# Contents

# Preface

When the publishers initially asked if we wanted to write yet another project management book, our first thought was that we had done the best we could already with earlier editions and had nothing more to add. After some thought, we began to see another audience for a stripped-down *guidelines* overview of the topic with a more prescriptive approach. The perceived audience for this version is individuals who are new to the project world with a need to get a reasonable perspective of management issues related to this activity without wading through voluminous levels of theory and mechanics. This could be a new project manager who wants quick help on key project management areas. So, our writing goal for this effort is to produce a tightly space-constrained explanation of the key management practices and steps that produce successful project results.

There are two driving beliefs that most influenced the covered topics. First, the average project manager has not spent sufficient time to know what research studies have shown regarding what causes projects to fail. This essentially focuses attention on the management process areas related to these factors. Second, we believe that an audience is less interested in theory and more in a more practical set of process guidelines. This approach allowed us to offer more personal experience than might be found in typical theory or tools efforts. Collectively, the authors have approximately 80 years of industry exposure associated with this subject, so we feel confident in the material shown here.

When one looks at the current state of project management across all industries, it would be fair to say that a significant proportion of projects suffer from a management maturity approach. Based on industry research and our knowledge, this text will describe a core set of management processes that should be followed. Realize that this compendium is not the universe of such processes but rather the core areas that contribute most to successful outcomes. Based on this discussion, our goal was not to deviate too far from a recognized and validated model structure and process.

As with most authors, ego pushes us to share everything we have acquired through the years. However, this is a tightly constrained scope effort in which we try to show an approach for dealing with projects using minimal descriptions as a quick read. Some personal bias topics did not fit the prioritization goal structure and had to be either left out or at least only minimally described. For example, the agile methodology is only briefly mentioned in later chapters, and Earned Value is only briefly mentioned. These are legitimate topics but are too broad to fit our space constraints despite being viewed as legitimate base processes. In an attempt to not alienate cultists from these two camps, we'll also politely say that neither of these methods guarantees a project's success without combining many of the basic topics outlined here.

Over our years of involvement in projects, we have experienced both success and failure. Later in our careers, as academics, we have tried to see what others have said about their methods and mapped this input to our experiences. From this effort, we conclude that managing projects of almost any type is very complex, and there is no single best way to execute a project.

Today's most visible philosophical conflict seems to be based on the approach related to (a) producing a clear definition of requirements before moving forward versus (b) doing less upfront definition and working through a series of small steps to evaluate what the product will look like as users evaluate early segments. Both approaches are legitimate management models depending upon the project type, but our bias is toward a middle hybrid approach where the project structure takes selected processes from each mode to achieve the best of both.

Today's project world is a hurry-up-oriented environment, and every project manager needs to be sensitive to this. To satisfy this need, it is necessary to execute the life cycle as quickly as possible. Spending inordinate periods carefully analyzing and planning everything will be reviewed critically. This is called *analysis paralysis*. The critical point that every project manager needs to understand is that *a project does not deliver its goal value until it is finished*. The iterative school claims an advantage because its incremental approach deals best with this delivery goal.

Conversely, it is equally important to realize that failure to deliver a flawed product can waste significant resources and competitive advantage. The best management approach is to find the right balance between failure from moving too fast versus failure from late delivery from overplanning. Ideally, the best method is to do the planning necessary to ensure successful completion. One comment frequently mentioned in the text is the value of using technology and templates to speed up the process rather than omitting it to save time. This approach would help keep the right process in place and still achieve the speed needed. One of the common process omissions whose value is often not understood is good communication. Research studies indicate that the primary reason for project failure is poor communication, which needs to be better understood.

We hope you find that this reduced-size format guideline overview delivers value to your goal of understanding the project world and its key processes.

*Gary Richardson*
*Deborah Sater Carstens*

# Authors

**Gary L. Richardson**, recently retired as the PMI Houston Endowed Professor of Project Management at the University of Houston, College of Technology graduate-level project management program. Gary comes from a broad professional background, including industry, consulting, government, and academia. After graduating from college with a basketball scholarship, he served as an officer in the U.S. Air Force, leaving after four years of service with the rank of Captain. He followed this as a manufacturing engineer at Texas Instruments in the Government Products Division. Later, non-academic experience involved various consulting-oriented assignments in Washington, DC, for the Defense Communications Agency, Department of Labor, and the U.S. Air Force (Pentagon). A large segment of his later professional career was spent in Houston, TX, with Texaco, Star Enterprise (Texaco/Aramco joint venture), and Service Corporation International in various senior IT- and CIO-level management positions. Interspersed through these industry stints, he was a tenured professor at Texas A&M and the University of South Florida, along with adjunct professor stints at two other universities. He joined the University of Houston in 2003 to create their project management graduate program and retired from that university in 2019. During this period, he taught various external project management courses to both international and U.S. audiences. He previously held professional certifications as Project Management Professional (PMP), Professional Engineer (PE), and Earned Value Management (EVM). Throughout his career, he published nine computer- and management-related textbooks and numerous technical articles.

Gary earned his B.S. in Mechanical Engineering from the Louisiana Tech, an AFIT post-graduate program in Meteorology at the University of Texas, an M.S. in Engineering Management from the University of Alaska, and a Ph.D. in Business Administration from the University of North Texas. His broad experience, associated with over 100 significant projects of various types over more than a 50-year period, has provided a wealth of background in this area as he observed project outcomes and various management techniques that have evolved over this time.

**Deborah Sater Carstens**, PMP, is Professor of Aviation Human Factors and Graduate Program Chair at the Florida Institute of Technology (Florida Tech), College of Aeronautics, USA. She is also the Florida Tech Site Director for the Federal Aviation Administration (FAA) Partnership to Enhance General

Aviation Safety, Accessibility and Sustainability (PEGASAS) Center of Excellence. She has worked for Florida Tech since 2003, where she instructs both graduate and undergraduate students teaching in on-campus and online programs. Formerly, she was a faculty member in information systems and the Academic Chair for the Project Management Track in the Online MBA Program when she worked in the Bisk College of Business.

Before Florida Tech, she worked for the National Aeronautics and Space Administration (NASA) at the Kennedy Space Center (KSC) from 1992 to 2003. In her career, both at Florida Tech and at NASA KSC, she has been the principal investigator of funded research from the Federal Aviation Administration (FAA), NASA, Small Business Administration (SBA), Florida Department of Education (DOE), Department of Health and Human Services (DHHS), NASA, Oak Ridge Associated Universities (ORAU), and other organizations. Her research has been in project management, pilot safety, human factors, learning effectiveness, and government accountability.

She received her Ph.D. in Industrial Engineering and B.S. in Business Administration from the University of Central Florida. She also holds an MBA from Florida Tech. She has over 75 publications.

# Introduction

# 1

This chapter outlines the design rationale for the text. The goal of the material selected represents the core processes that have a significant correlation to impact project success. One might ask why read this text when there are so many others in the marketplace? The answer we offer is that this structure offers the best exposure to the topic in a limited space. This is for an audience that doesn't want or have time to read massive texts on the topic. Simply stated, we believe that this format fits an audience that likes a condensed review, much like the old *Reader's Digest*. This will be true, especially if the content selected does present key topics that potentially impact project success. In other words, give the reader time-tested advice and basic mechanics on project areas that are critical to success. Admittedly, this format will not contain everything you need to know, but it does offer the biggest bang for the time commitment. That is the essence of this text goal—quick answers to key topics with links to more detailed sources.

Each of the chapters outlines the general management activities related to that process. Space constraints limit explanation of related software mechanics or extensive documentation artifacts, but both of these areas will have reference linkages to external sources that would help with an expanded view if the reader is inclined. The clear goal is to offer usable descriptions of project management in a minimal format.

A common industry says, "it is hard to think about theory when you are trying to drain the swamp and alligators are biting you in the rear." That is the way it feels with project management. The day-to-day issues can be so hectic that focusing on dealing with any one thing is hard to do. Some of the topics described here might be looked at as obvious, but experience suggests that these chapter topics are the ones not being done well or not at all. In any case, shorting any of these topics compromises the project manager's ability to guide the outcome properly and may make the desired result infeasible. Simply said, these topics are the key to success. Some of the topics described have been tightly compressed, but the goal was to leave the main content intact. This topic collection does not mean that other topics are unimportant, just that the others do not impact the outcome as much as the covered topics do (in

DOI: 10.1201/9781003218982-1

the author's judgment). Also, because of space constraints, mechanical key-stroking details are removed and linked to outside resources. The referenced areas represent the fringe of the typical reader's needs and the outside resource details are easily found. There will be some experts who say that we have missed a topic or two. That is the nature of this subject, but we feel comfortable with this limited selection as a worthy set to understand.

The thesis of this content selection is that it includes approximately 75% of the root cause of project failure. Does this mean that the project will be 75% better or even successful if these guidelines are followed? Unfortunately, no written description of that topic is that good. There are so many operational variables in a project that prohibits distilling management guidelines into cookbook checklist formats. However, handling these book topics correctly will cut down the likelihood of major variances resulting from these areas. Another educational source of information for this topic is the daily news. Major failed projects show up there frequently, and these are being executed by professionals. Theoretically, reviewing real-world case studies and comparing that material to this text will provide some motivation that this set of material is legitimate. In other words, what did the real-world project not do properly? It is easy to second guess such decisions, but you will see a process that could have been executed better in most cases.

Literature related to project management is a popular topic, and each text or article offers some new solution to a problem area. Individual views outlining the best approach often represent the authors' opinion. Some authors approach this problem by laying out a full theory book supported by some international support organizations such as the Project Management Institute (PMI). One of your authors has done that under the title of *Project Management Theory and Practice* (Richardson and Jackson, 2019). A second writing approach illustrates how various supporting tools such as Microsoft Project can handle project management activities. Your authors also executed that model with *Project Management Tools and Techniques* (Carstens and Richardson, 2020). Both of these efforts are valid documentation of those two approaches, but each requires a significant time commitment to absorb. Yet, a third approach to obtaining usable data on this topic comes from the myriad of authors who tout their unique approach, usually on some isolated management process. For example, the agile school of project management is quite popular these days as it is often touted as the best method to achieve success. In addition to these, there are many other sources to document various aspects of this complex area. Frankly, there is some merit in all of these contributions, and all seek the holy grail that would solve the problem of managing this class of work. In some metaphoric way, this collection has the feel of the Big Bang theory. There is a lot we don't know, but a lot we think we know. Just realize that there is no simple answer that everyone agrees with. If you are a curious reader of this subject area, you need to continue to explore these various offerings in hopes that some source will provide you with useful insights. Until you choose to do this, we believe this collection of the material described here will provide a good starting place with minimal effort.

Projects today still have significant failures, and surprisingly the same root problem areas seem to repeat with great frequency. Even seasoned project managers are not always successful. More details on failure statistics are described in Chapter 4. As indicated earlier, project environments are complex (just like the Big Bang theory mentioned above). Project executive sponsors, users, and other stakeholders have differing views on what to do and how long it should take to finish. The project manager is caught in the middle of the maze. Even the definition of success is not clear in many cases. For example, is the primary goal to produce a specified gold-plated widget, or just a plain cheap one (surely not a gold-plated and cheap one)? What about the desired schedule or budget? Yes, hundreds of things can go wrong in the project life cycle, but trying to learn and deal with these seems to be overwhelming even with experience. The thought here is that if you only take time to learn how to use a hammer, a project looks like a nail, but this topic is much more understandable if one takes the time to look at a general life cycle model and its associated processes. The theme espoused here is to say, "focus on the main process areas for sure and then work outward to others as you learn more about the correct process in your particular environment."

Here is one other often misunderstood point. There is a lot more commonality in various project types than there are differences. The goals of each particular project will be different, but the core processes remain quite similar. For example, making a movie has many of the same processes as building a skyrise building. One can even argue that your personal life should be managed like a good project with plans, phases, risk assessment, etc. The management model for these various project types is generally the same; at the core are ten major management areas consolidated into this book and associated key processes that need to be addressed. It will be up to the project manager to decide exactly how each of these processes will be utilized and what level of structure and formality is justified.

Think of this text content design as an upside-down approach. Rather than looking at the universe and trying to find Earth, we start with Earth and look outward only as necessary. You may not need to know all about the rings of Saturn, but you might want to know more about local rain forecasting. Assuming you don't have the time or interest to spend looking for a detailed discourse on the broad topic, it is hoped that this will provide what you need. Humans are good problem-solvers once they know the problem to be solved. The collection of material shown here represents that list.

Since every project is different, appropriate processes to deal with these differences have to be customized so that any theory will need that treatment. After being involved with the project environment over many years, we see organizational approaches that do not reflect this idea (i.e., how to construct a project plan that has integrity, how to track status, what to measure, how to resolve conflicts, etc.). Also, how to manage project resources and scope creep, etc. The list goes on. So, if we claim to know what key areas to deal with, how do we describe solutions to such a broad area? The first part of each chapter outlines a description of the reason that topic is important. From that, more mechanical management-oriented descriptions will be given. Each topic will normally be of 8–12 pages in length. If more explanation is desired, there will be sources mentioned that will provide additional information. Readers who are new to the project management world or those looking for an abbreviated quick read view should find this material and approach appropriate.

Now to writing style. An effort is made in each section to explain in plain language why the target topic creates issues. From this, simple examples are given to help further describe that area. Recognize that using these examples will not make the problem go away. However, it will help identify key processes to focus on. Think of this as a roadmap of key project processes. The hypothesis is if you don't deal with this topic in your project, a lower than desired result will follow. One might say, "if all of this is so obvious, why is it not written in stone for the project manager?" The answer to this question is not so clear. Yes, these subjects are known but still not handled well in many cases. In this approach, you will find a roadmap showing major cities (processes). Purists will say that this approach is too high level and are even naive to suggest that this represents 75% of the failure sources. So be it! If you, the reader, get efficient reference value out of it, we will take that criticism. One can always go to various reference sources and get more details if that proves your particular need.

# REFERENCE

Richardson, G. L., and Jackson, B.M. 2019. Project Management Theory and Practice, 3rd ed. CRC Press, New York.

# Project Environment

# 2

---

## INTRODUCTION

---

Think of a project as an activity to change an existing product or process or create a new one from a vaguely defined vision. This aggregation of visionaries, technical specialists, management, and future users represents a complex activity that occupies many organizational resources. Some of these ventures bring great competitive advantage, while others make headlines with product failure, budget overruns, or other visible negative outcomes. The purpose of this book is to explore this organizational area in an abbreviated form. Assuming the project gets through the formal portfolio review, it enters a complex series of processes designed to complete the task. Step one of that approval is to formally assign a project manager who will have the primary management responsibility for the upcoming effort. Management should also guide the high-level goal outcome targets, including schedule and budget.

---

## PROJECT VISION

---

One of the more interesting aspects of a project is the environment in which it first emerges. There is no singular view of this, but the essence is a mentor who tries to birth his vision. This can be a new product, a new way of doing business, or simply upgrading an existing version. Regardless, to move forward, there is a requirement to obtain some consensus that this idea has merit. There is certainly a selling aspect to this activity, and mature organizations require that a formal business case be prepared to quantify the perceived value. This will be described later regarding how this document evolves.

A second view of the vision process is to evaluate how it fits into other projects. Each project competes for organization resources. Large organizations

DOI: 10.1201/9781003218982-2

will have hundreds, or thousands of projects proposed and in work at any one time. These projects compete for the same resources in terms of money, human skills, and even facilities. As a result of this, there is heavy competition to get a defined project approved. The common term for this is portfolio management, and just like a stock market review, only so many stocks (projects) can be approved based on resource constraints.

# THE PROJECT ENVIRONMENT

Assuming the project gets through the formal portfolio review, it enters a complex series of processes designed to complete the task. Step one of that approval should be to formally assign a project manager who will have the primary management responsibility for the upcoming effort. Management should also guide the high-level goal targets for the outcome, including schedule, budget, and human and technical resources. The critical thing to understand is that the actual feasibility of executing this goal has not yet been determined. That is the goal of subsequent planning tasks. Recognize that a project goes through an evolutionary set of phases, each designed to move forward toward the end goal and then terminate. This is called the project life cycle—a birth to death view. It may help to envision this process by thinking of each step as a complex movement of resources in and out of the effort, with each of these steps having a goal. In the early phases, the focus is on deciding what is technically feasible and then later matching that with the needed resources to accomplish that scope of work. This is called the planning phase. At the end of the planning effort, the project structure has been established and also the associated goals. Questions related to what, when, who, risks, and how much are now quantified. From this, senior management has the responsibility to approve the plan and its related goals. During this evolution, the resulting plan likely will not match the earlier business case as more is now understood regarding the underlying factors.

In most projects, there is a requirement to create a project technical team to focus on the development activities. One of the initial complexities is often to extract the needed resources from other organizations. These resources may be loaned to the team or permanently assigned, but this acquisition process is similar to drafting a sports team. The project manager is looking for all-stars, but the journeyman may be all that is available. Finding enough team members with the appropriate skill is the first management complexity to be overcome before the process can begin.

Surprising to many, the original vision is seldom adequate for a technical specialist to produce the desired outcome. This understanding gap between visionaries and the construction process is the first major risk point for the project. Working out these understanding gaps will involve a heterogeneous collection of individuals with diverse objectives. The visionary side is interested in the functionality of the target. Technical specialists are more interested in how to construct that goal. Management and financial stakeholders focus on resource aspects. Collectively, the goal is to piece together a roadmap that defines product functionality, schedules, budgets, and resources related to the effort. As this collective view of the project evolves out of the definitional fog, risks and constraints emerge. Desired product performance can't be reached given the cost budget, resource availability, or schedule constraints. Risk levels are identified, and these may violate constraints as well. Out of this fog must come a roadmap to achieve some agreed-upon output goal set.

Chapters 3, 4, and 5 address different perspectives of this activity. This includes management roles, project drivers, and industry success data. Chapter 6 will describe how projects evolve their work definition, which will be defined as scope. Chapter 7 describes how to construct a project schedule from the defined scope. Chapter 8 describes how to develop a project budget from the scope definition. Chapter 9 describes how to assess and deal with project risk.

## EXTERNAL ENTITIES

External entities are individuals or organizations that are external to the project. These entities can affect the project in some way. For instance, an external entity could be a supplier. If the supplier incurred a transportation delay or a delay in creating the item ordered by the project, a planned deliverable could be delayed. If one deliverable is delayed, this could cause additional delays for other deliverables, and ultimately the entire project end date could be delayed because of the external supplier. Another example of an external entity is a public environmental group. Their opposition to the project goals could create negative publicity that might cause the project to be shut down. Also, the surrounding community could have similar objections that need to be addressed. A reason for negative publicity could be the potential loss of local jobs related to the new project. On the other hand, positive publicity from a newspaper article about the project team working on some volunteer effort would be desirable. There are various relationship links to external entities that the project needs to be aware of in order to manage the project accordingly.

# PROJECT EXECUTION

The model assumption for project execution is that it is guided by management's formally approved plan. This means, in theory, that "all" the team must do is produce the defined work units in the defined sequence, and the result will be as planned. Unfortunately, that is not the reality because actual task times are typically larger than the initial estimate, schedules and budgets can be overrun, and unexpected risk events can emerge. Furthermore, resource gap issues are prevalent in terms of both quantity and quality. In the best case, one could describe this environment as "dynamic," while it might be best described as pure chaos. Regardless of the actual state, there are various "events" parts to deal with during the execution process, and the project manager's skills will be challenged.

Some also believe that the essence of project management is to collect some resources, and they will automatically lead the effort to succeed. That is not the normal result for sure! It takes knowledge of the appropriate processes and management skills to execute this activity. There is now sufficient history and research related to this activity to define the necessary activities to maximize success. The caution here is that the potential variables involved are great, and some projects will not be successful because the environment does not match the goal set. It is possible to describe the positive side of these variables, but it is not necessarily always possible to manage the negative side.

# DEALING WITH UNPLANNED VARIATIONS

Recognition that unplanned variations will occur necessitates actions to identify the variation and various management activities to correct the root cause. This requires the measurement of actual results as a starting point. Future chapters deal with various aspects of the ongoing management processes. Chapter 10 describes the communication goals needed. Chapter 11 describes some of the activities used to speed up a lagging project. Chapter 12 describes the activities involved in monitoring and control of the key elements, and Chapter 13 discusses the management of project team members. Techniques to handle plan variations in the plan are scattered throughout the book. Because of these variations, some believe that planning is a waste of time. However, the authors hope that this book will provide theoretical and mechanical guidance to project managers regarding working through unplanned events to fix whatever emerges.

# WRAP-UP CHAPTERS

The final three chapters outline three additional areas that need to be understood as part of the final overview. Chapter 14 offers a brief discussion of closing the project, which is an often-ignored process. Chapter 15 essentially opens Pandora's box to discuss the contemporary agile iterative (prototyping) methodology, which refutes much of the processes contained in this book. There is certainly much to be learned from the agile industry experience, and these positives will be described here. Chapter 16 closes the textbook by offering a view of project management in the future. It outlines key concepts that need to be employed in all life cycles regardless of whether the design is a traditional waterfall or an iterative form.

# SUMMARY

This chapter has addressed the general process by which the project is envisioned, approved, and planned. The perspective of a competitive organizational portfolio analysis of project proposals is mentioned. When the project is selected to be undertaken, it is validated as a target goal with perceived value and alignment with organizational objectives. This brief introduction summarized the project life cycle phases.

A key point is made, that projects do not naturally evolve in an orderly fashion according to a predefined plan for several reasons. These unplanned variations require management knowledge and skill to correct them. There are three primary goal outcomes for the project: planned function, schedule, and budget. These three variables interact, and sometimes it is necessary to trade off one of the variables for another. For instance, additional resources can cut the schedule or expand functionality. Likewise, cutting functionality can save time or budget. Project plans establish targets for all three of these primary outcome categories. Future chapters will describe how these variables are defined and tracked throughout the life cycle.

# Project Life Cycle Drivers

# 3

---

## INTRODUCTION

---

This chapter outlines the high-level drivers that shape the character of a project and help determine the proper steps to navigate through them. All projects are different, and it is necessary to understand the role of these variables and the associated management process. There are many "drivers" that determine the character of a project. If one can accurately assess these drivers, it will help in structuring proper management for the project. This section will attempt to build a coherent view of a project by looking at four aspects of the underlying architecture:

Product factors
Demand variables
Supply-side factors
Support assets

The project manager needs to assess these four groupings regarding their status and impact on the project life cycle. The story below will paint a picture of these management factors as they affect the life cycles. Some of these drivers shape the project goals, while others impact the input or output variables.

---

## PRODUCT FACTORS

---

The highest-level view of the project is the essence of what is the desired output. If we are building several similar houses in a subdivision, this has a lot to say about our ability to estimate the effort accurately. However, suppose we are planning a SpaceX rocket to blast off and return to the landing pad. In this case, it is quite another project in that accuracy of almost all variables is less than the standard building type project. R&D-type projects do not fit

DOI: 10.1201/9781003218982-3

very well into the management model outlined here. The model described in this test starts with a goal that can be defined and a technical pathway that is understood. The technical model with a defined goal and pathway is the model driving factors that are outlined here.

# DEMAND-SIDE DRIVERS

When a project is initially envisioned, the focus is on the desired functional output. This involves working with multiple organizational groups to negotiate a balance of planned scope, schedule, and budget for the activity. A more detailed planning process will attempt to answer general feasibility, realistic time frame, cost, risk, or other factors that affect successful delivery. In some cases, the product's value is time-sensitive, and the scope defined may not be feasible in that time frame, so the project can be done but not when needed. The value is zero if that date is missed.

Before the vision discussion goes very far, some organizational decision authority level must be convinced about its merit. At the early stages, major discussions wrestle with what is envisioned, when it is needed, and what perceived value it has. Management approval from this analysis step allows the process to commence. Lastly, high-level discussions of the related risk would be undertaken. This descriptive data is packaged into a formal business case document for external review. If the factors described are judged to have both technical and organizational merits, the project would be approved through a process that varies across organizations. In our model theory, a formal project Charter would be created to show that the project is approved by senior management officially. Some funds are approved, and resources are assigned to support the formal planning process.

Even though the project has been approved, many necessary decisions remain, but these will have to wait on a clearer definition of the requirements and estimated work effort. The initial management approval frees the process to move on to the next step and refine the high-level Business Case data. This step is called the planning process. The first action here is to refine the technical scope of the effort (more in Chapter 6). The operational theory behind this refinement step is that you cannot accurately define the schedule, budget, or risk of a project until you have attempted to define the work required. Scope definition lays the foundation of the project on which all other parts must fit.

Based on the refined scope definition, work units are defined, estimated, and budgeted. These steps will result in what we call a first cut schedule and budget. From this base definition, project risk factors are identified and analyzed. Reserves are then introduced to cover variations in schedule, budget, risk, and potential changes in scope (user requests).

To have a valid plan, the project owner must agree to the newly refined demand variables produced through the planning phase. At this point, the Business Case values could have been off by 100% as more becomes understood about the effort. Management might require additional replanning to fit some constraints—usually schedule or budget. There is a fancy term used here that is not readily understood. This term is called *Progressive Elaboration.* One view of this is to think of a project as an onion. As the layers are peeled off, new insights are uncovered. Note that the Business Case was elaborated through the following planning activity to create a more viable scope description. This elaboration concept is an important idea in the evolution of a project. New layers of understanding are created at each stage, but these new insights must remain linked and integrated into the original vision. This concept remains all through the planning cycle and beyond as new layers get exposed. After what may require several iterations of planning, rejection, and finally, approval, all project puzzle parts are judged to fit (i.e., scope, schedule, budget, risk, and resource availability). These decisions are then packaged into an approved project plan, and that document defines the demand side of the project, showing what is to be accomplished.

The planning description above fits closely to what is called the *waterfall model* of planning. The plan produced conceptually represents the optimum way to achieve the results. However, if during this exercise, the goal is to produce this output quicker than that, many of the elaboration process steps will be done differently or not at all. Chapter 11 describes some of the activities that may be undertaken to shorten the life cycle, and various of these shortcuts may be used in the normal process. Recognize that the goal of this chapter and the entire text is to describe a desirable management approach. No, you don't have to plan, nor do you have to evaluate risk or availability of resources. But when a decision is made to shortcut these model steps, discussions should be undertaken to evaluate the negative side of that decision. When any one of these areas is not as assumed, the project will take a hit. No matter what development decisions are made, project success will be measured primarily using defined measures of the product performance variables.

## SUPPLY-SIDE DRIVERS

The supply-side drivers represent the array of resources and processes that collectively produce the project goal. The human element includes both the quantity and skill levels required. Before a plan can be approved, a match of available resources to the plan requirements is needed. This process is called *Capacity Management*. One does not have to ponder long to understand that failure to supply appropriate resources to tasks leads directly to a schedule overrun. Balancing these two elements is paramount to success.

The second view of resources is the project team structure. This has many characteristics of a sports team (i.e., different skill levels, morale issues, conflicts, etc.). Some project teams learn to work together and become what is called a *High-Performance Team* capable of working without a lot of direct supervision. Other situations are not so good. It is up to the project manager to ensure that the product objectives are properly communicated to the delivery team. Also, there is a motivation aspect to that goal. In many project environments, the resources are borrowed from other organizations, leaving the manager with less authority over the team than he would like. The project manager has a pivotal role in leading the team and organization through a complex collection of steps and events to complete the life cycle.

The project development process description outlined here is heavily supply-side oriented, in that the focus is more on managing the various supply processes to achieve the goal. Once the project enters the execution phase, the plan begins to incur deviations, and the management process involves significant effort to move the results back toward the desired product goal. More specific details on the management of variation sources come later. Much of the text will focus on activities which the supply side resources will take to correct plan variations.

# SUPPORT ASSET DRIVERS

The final component of the project driver variable set involves external entities that play different roles. The most important member of this group is the senior manager who owns the project (per organizational theory). In a pure model view, the project budget resides in his management purview. As a result, his authority represents an important role in the overall management scheme. For example, if the project status variables go outside of approved bounds, this person can decide to cancel the project. Maybe a memory example will help you remember this point. Metaphorically, if you asked, "where does a big gorilla dance?" After reading this section, remember the answer to this is "wherever he wants." That is the attitude the project manager needs to understand regarding how he deals with his project owner. In reality, the world may not be so cruel (or clear), but it is best to assume that this individual certainly gets your best attention. This person needs to understand both why and what is going on if you are going to build their confidence in the team's performance.

A second support group to deal with is the project stakeholders, defined as any individuals involved or interested in the project. As an example, a future user of the output is a key stakeholder. Likewise, a governmental environmental regulatory group might be very interested in the project and could have the power to impact the outcome. Over the past few years, there has been growing recognition of this group's impact on the project. This means that they are a major communication source for project status and input. One pressure they typically bring is increasing the scope and decreasing time schedules. This paradox increases conflict in the team as they would prefer to hold the initially approved scope fixed. Scope changes tend to drain resources away from the defined project work and make any newly defined work more complicated. Another aspect of stakeholder dealing is in communicating with them, so surprises do not occur in the future. The worst statement you can hear from a key stakeholder is, "oh, I didn't know you were doing that (to my favorite requirement item)."

A third support domain is embedded in the overall organizational maturity (processes, culture, etc.). Some organizations are very mature, and most of what a project needs are readily available from that source (i.e., computers, labs, offices, accounting, technical support, project templates, administrative, etc.). Beyond the raw resource support, mature organizations have operational processes already in place for a project to follow. If these items are not available, it will be time-consuming for the project team to custom-build what is needed. This would also drain resources away from the core task.

A final example of a support driver comes from external environmental factors. This includes any external source that could affect the project (i.e., items such as financial markets, political rulings, environmental regulations, etc.). In some cases, these are generally uncontrollable factors, and in others, the source may need to be nurtured into a more cooperative relationship. Chapter 13 will provide more detail regarding the relationship issues with project stakeholders. For this introductory view, the point is to recognize that other groups outside of the tangible project organization can affect the outcome of the project both positively and negatively.

## SUMMARY

This chapter has described a high-level and somewhat theoretical overview of the main project drivers that have a role and impact on the subsequent project outcome. A project manager needs to assess these factors and decide how the project will best interface with each driver category. None of these can be ignored without incurring undue risk.

# Project Success Factors

# 4

---

## INTRODUCTION

---

The goal of this chapter is to establish key attributes related to project success and failure. Two industry trend data surveys will illustrate what various organizations are doing to improve their operational maturity.

The first item to resolve is the definition of "project success." Typically, success is measured by a custom combination of four variables—schedule, budget, functionality of the deliverable and possibly quality. There are various ways to measure project outcomes based on these measures, but simple measures such as those outlined above are not adequate given the inherent complexity of the project environment. Regardless of how one goes about measuring success, there is universal agreement that project outcomes are often not what one would like to see. In recent times, some mature organizations have started to claim improved success rates, but the integrity of these reports is often questioned.

To help in understanding current industry status, trends and performance results from the Project Management Institute (PMI) Pulse Survey and the Standish IT surveys are described below.

## PMI Pulse Survey

PMI performs an annual survey of project status reporting these results (PMI Pulse, 2017, p. 5):

- Projects hit the major business target 60–70% range—this does not mean it succeeded but was focused on the right target.
- Projects on average fail in the teens (approximately 18%).
- Projects finish on schedule and budget in the lower 50% range.

Over time, these trends do not show great improvement. Does this mean that organizations are not learning how to manage their projects better, or is there

DOI: 10.1201/9781003218982-4

some factor hidden below this level that inhibits improvement? The answer to that question is much more complex than a yes or no.

One result found in all surveys is the performance differences in large projects versus small ones. Some stated reasons for this are as follows:

- Smaller projects can better define requirements (smaller scope).
- Smaller projects have fewer tasks to manage (also less time to implement).
- Smaller team size makes communication easier—the Pulse survey divided the respondents into Champions and Underperformers to evaluate their differences in performance. One metric collected is waste (inefficiency), and it is estimated to be 12% of the total budget on average.
- Smaller projects require dealing with fewer stakeholders.

Each of these observations is important to understand in structuring the scope of a project.

The higher-performing Champions group wasted 1/28th as much as the lower Underperformer group. In addition, the higher performers enjoyed more successful business outcomes and fared better at other measures of project completion (PMI Pulse, 2017). Many organizations still assume that project management is learned through experience only, which is a questionable strategy. The Pulse Survey Champions Group performed better by focusing on the development of technical skills (76% versus only 19% for underperformers), leadership skills (76% versus 16% for underperformers), and strategic and business management skills (65% versus 14% for underperformers). Each of these critical areas was judged to affect project outcomes. These statistical comparisons indicate that developing an appropriate management culture is worth pursuing.

# STANDISH SURVEYS

The Standish Group has been performing analyses of IT projects for many years, and these results have shed considerable light on project performance (Lynch, 2015). Rather than the traditional three success metric factors, there are now six measures: schedule, budget, achieving the target, on goal, business value, and user satisfaction. Even with this expanded definitional version, the question remains on how to measure each parameter and when to do this (i.e., should the project end be initial production or some future point with more experience with it, etc.). This measurement ambiguity leaves somewhat of a dilemma regarding how to grade the end-of-project values. When it is all said and done, success measurement lies with the organization and the user community. If the result has more value than the project cost and the users feel that it was successful, should it be graded as successful? The measurement process is just one more item in the complex nature of project management.

One of the most significant factors that correlate with traditional success measures is project size. Although not a surprising revelation, it is important to recognize that larger initiatives are more complex in many ways, which is reflected in the results. What is not so obvious is the degree to which this statement is true. The success rate of small versus large projects is (82% versus 2%) in favor of small ones (Lynch, 2015). The failed category is not significantly different across the size groups, but the extreme differences on the success side must be kept in mind. To summarize this differently, the overall success of IT projects has been below 30% for at least the last seven years, and failure rates have remained around 20%.

The Standish surveys also reviewed the top ten factors that correlated with success (Lynch, 2015). These factors were also weighed on the overall impact, with 80% of the graded success impact found in the following five factors:

- Skilled resources (20%)—Of all the factors on the success list, this is the most obvious. There is a clear need for skilled human resources that understand how to execute complex technical undertakings.
- Executive sponsorship (15%)—Support from senior management.
- User involvement (15%)—Assists in properly defining the scope and supporting project needs throughout the life cycle.
- Emotional maturity (15%)—This is a somewhat catch-all phrase dealing with the soft skill maturity of the organization (i.e., managing conflict, handling motivation, etc.).
- Optimization (15%)—This term is one of the more difficult factors to define, but the essence of this lies in how the project is managed (i.e., business focus, internal processes used to achieve the goal, etc.).

# CONTEMPORARY TRENDS

There are various initiatives underway across the industry to support higher success rates. We have selected two major examples to show here—PMI's

Pulse (2017) listing and American National Standards Institute (ANSI) standard process model for Best Practices (AQNotes, 2018). The first example is an industry trend of the most used management processes, while the second is a well-respected integrated management process standard. Numerous other agencies describe a model environment, but these two samples provide a reasonable overview sample to illustrate key management processes. The chapters in this textbook highlight the authors' condensed view of standard management practices, slanted toward a practical implementation approach.

## PMI Pulse Management Practices Trend

The PMI Pulse (2017) survey tracked trends of 11 key management processes. This would at least suggest that these management actions were believed to be productive to increase success. Each of the positive trends correlates with various model process areas discussed in various sections of this textbook. The key process areas are as follows:

- Change Management*—mentioned in Chapter 12
- Risk Management*—described in Chapter 9
- Having a PMO—evaluating global project need by priority; not covered in the textbook
- Using Program Management*—grouping projects into program groups; not covered in the textbook
- Standardized Project Management**—a major topic of the textbook
- Portfolio Management—viewing the overall collection of projects in the organization; not covered in the textbook
- Focus on developing Project Manager Competency—a declining trend, not covered in the textbook
- Using Earned Value*—sophisticated tracking measure mentioned briefly in Chapter 12
- Use of Agile Model Methods*—technique to speed up completion that is growing in usage; mentioned in Chapters 1, 15, and 16
- Use of Extreme Model Methods*—not mentioned in the textbook

The items with one asterisk (*) indicate that the process is always or often used, while the one item with two asterisks (**) indicates that it is used often or always across all departments. This list does not necessarily equate to a success strategy listing, but it does show what organizations are doing to be more successful.

## ANSI EIA-748 Organizational Maturity Specification

An operative definition of "environmental maturity" comes from the industry-standard ANSI-EIA-748 (AQNotes, 2018). This set of operational processes now includes five major groups and 32 embedded goal-oriented criteria. The five major process groups are as follows:

A. Organization—oriented toward various structural-type components
B. Planning, Scheduling, and Budgeting—a close mirror to Chapters 5, 6, and 7
C. Accounting Considerations—process definition for collecting, recording, and tracking project costs
D. Analysis and Management Reports—techniques to evaluate and report project performance, present, and future, emphasizing earned value metrics
E. Revisions and Data Maintenance—processes focused on managing the project baseline and change control.

These process guidelines are used to validate the project infrastructure of most government contractors.

This specification consists of 32 named processes under the five categories. It is important to note that these processes have been evaluated over some 40 years and represent an integrated view of a mature project environment. A detailed review of these guidelines would provide a strategic map to improve local processes; this standard is available at: https://www.dcma.mil/Portals/31/Documents/Policy/DCMA-PAM-200-1.pdf?ver=2016-12-28-125801-627

---

# SUMMARY

---

This chapter has provided a brief overview of the success and failure characteristics of projects. Two survey views have been used to illustrate how this topic is viewed in practice. A standard model project environment is summarized with the ANSI-EIA-748 guidelines, and an external reference is given to explore that in more detail.

All the chapters in this textbook are geared to providing good practice guidance regarding a strategy for achieving successful project outcomes. The textbook chapter material collectively is an abbreviated approach to defining a practical view of a successful model.

# REFERENCES

AQNotes. 2018. ANSI EIA-748 earned value management. Defense Acquisition Agency. Available at: http://acqnotes.com (accessed January 15, 2019).

Lynch, J. 2015. Standish group 2015 chaos report—Q&A with Jennifer Lynch. October 4, 2015. Available at: https://www.infoq.com/articles/standish-chaos-2015 (accessed November 4, 2017).

PMI Pulse. 2017. *Pulse of the Profession*. Newtown Square, PA: Project Management Institute.

# Project Manager Roles and Management Structure

5

## INTRODUCTION

This chapter discusses the power of the project manager and different organizational structures, each having inherent different degrees of power that link either to the project manager or to the functional manager. The five different organizational structures that will be discussed include functional, matrix, virtual, project-oriented, and hybrid (PMI, 2017). Each structure has distinct characteristics, benefits, and challenges.

## POWER

Projects are natural breeding grounds for conflict, which often leads to ineffective human behavior. These situations must be dealt with effectively for a project's overall success and health (Richardson and Jackson, 2019). Power can be in terms of the project's senior management support and commitment to bringing it to fruition (Maddalena, 2012). These struggles stem from a limited supply of resources such as labor, monetary, facilities (space), and equipment. Everyone is competing for the best employees, project dollars, best space, and best equipment. With this in mind, executives spend time planning which efforts or projects their company should invest in, resulting in the maximum growth of a company's profits. Just like a sports coach wants to recruit the best players, the project manager wants to recruit the best project team members. All

DOI: 10.1201/9781003218982-5

of this becomes a political game because limited resources result in unlimited power struggles within an organization at any given time. Different communication channels go up and down the organization as employees ask their bosses for monetary resources such as funds to upgrade equipment. In return, a specific organizational level then asks their managers for funding for the employees' needs, resulting in managers asking their bosses and their bosses asking the company's executives for funding. This chain of demands permeates the organizational structure. Different employees, leads, managers, and executives continuously have to make choices regarding the best way to use their limited monetary resources to address the endless funding requests. Sometimes, the choices made regarding how to spend the funds have to do with what is best for a company, but often choices have less to do with a company's needs and more to do with funding the manager with the greatest degree of power.

There can also be hidden agendas at play, such as a business unit creating a need for a specific project for their own value without aligning their project with organizational-level goals. These hidden agendas have more to do with returning favors to individuals to keep conflicts to a minimum. Obviously, this is never the recommended approach to leadership and management. Managers might feel the pressure to provide funding for one project over another based on a favored personal relationship or promised favors in the future. Ideally, decisions on how to use the limited amount of funding should be based on which projects or needs are best in global alignment with organization goals. However, in reality, this is a true challenge because of the interpersonal factors outlined above. Another problematic decision topic area is that of outsourcing. The decision to outsource is often made purely based on current cost, and the middle segments of the organization are typically most impacted by this decision. Later, the organization may find that the strategic value of the decision to outsource was not as anticipated. This section touched on some of the typical power challenges within an organization.

# FUNCTIONAL ORGANIZATION STRUCTURE

The functional organization structure is "an organizational structure in which staff is grouped by areas of specialization, and the project manager has limited authority to assign work and apply resources" (PMI, 2017, p. 707). The term "functional" is the same as "centralized," according to PMI (2017, p. 47). According to the major business functions, this structure is set up with examples displayed in both Figures 5.1 and 5.2. Therefore, the different departments of divisions are formed based on the company's main functions, and employees have one clear supervisor. A benefit of this structure is that people in a specialty unit have an in-depth understanding of the functional area to which they belong, such as aircraft-related projects. These individuals are the subject matter experts (SMEs) for that function. Suppose there is a project that involves the development of a system to support a particular function. In that case, SMEs from that functional unit can develop system requirements so that the system deliverable involves the user's perspective resulting in a useful system for that functional unit.

**FIGURE 5.1**  Functional organizational structure example 1.

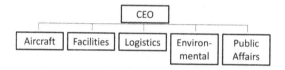

**FIGURE 5.2**  Functional organizational structure example 2.

A benefit of the functional organization structure is that working relationships tend to be more established and clearly defined. This is due to a fixed contingent of people on a team knowing how to successfully work with each other and managers already knowing the strengths and weaknesses of different team members. This occurs because these individuals stay in their assigned roles for fairly long periods. Suppose one person, for instance, is very good at researching information and another one at presenting it to customers. In that case, the boss can use this knowledge to have a very successful meeting with the customer—full of detailed information yet presented by an individual with excellent presenting and networking skills. Knowledge of the strengths and weaknesses of different individuals and their capabilities make for selecting the project team better. Therefore, the director can decide which individuals are best for different project assignments. This only further contributes to successful work efforts.

Within a functional organization structure, the internal work units do not have to compete with one another to support their specialty groups because only one group is familiar with a particular work activity, and this eliminates competition within the same organization. With less competition, there are fewer power struggles and, therefore, politics. This is not to say that politics and power struggles are not occurring at the upper levels, where directors from each functional unit compete for the company's limited resources. However, there are at least fewer politics within the lower and middle levels of the organization, which is a true benefit. Another benefit is that functional groups are often very adept at technical problem-solving within the group because all their projects are related, and, thus, expertise is acquired by the group. Therefore, individuals within the functional unit have extreme knowledge within their field, so time is not spent getting each team member up to speed on the focus of the project. Instead, the project team already understands specific challenges that exist within their function to be very efficient and successful at problem-solving.

There is also a more straightforward line of authority within the departmental organization within the functional organization structure. This linkage structure makes it easier for priorities to be set and conflicts to be resolved. So the focus for a work team can be set quickly, and any conflicts resolved quicker, allowing for the team to get back on track toward the project deliverables. When there is a clear line of authority, it is easier for employees to have less conflict because everyone is aware of defined roles and responsibilities when something is not being handled appropriately. In this stable environment, work teams can easily be held responsible and accountable for the success of the defined goals. Therefore, clear authority increases pressure for people to honor their commitments because performance appraisals, for instance, are conducted by the same organization where the work is being accomplished. Furthermore, clearly defined career paths exist for people who do good work as their supervisors notice them and reward them through incentive pay or promotions.

A functional organization structure has a very clean way of exercising control over its decision structure. This occurs because all requests are submitted to a group manager for approval, including identifying the team members to work on a project. In this functional organization, there is typically one set of management procedures and reporting systems for all work efforts within each functional department, making the decision processes clear. This is a large benefit to the company having projects that utilize team members from one functional department because these procedures and reporting systems should be more perfected than each project manager creating their own procedure and reporting system. In this structure, it is easier for higher-level managers to see the health of their functional department to include weak procedures and reporting systems. In many ways, each functional department is like an island with a well-defined chief who tends to run that segment as though it was the entire organization. This begins the discussion on the challenges of projects that pull from the different functional departments because each has its own processes.

Each functional unit acts almost independently and most likely has different work procedures and reporting systems. Therefore, the company, as a whole, has inconsistent procedures and reporting processes. For example, a project that pulls from more than one functional division results in higher-level management not being able to compare apples with apples when assessing the organization's health because it will be more difficult to compare one functional unit to another. This is because all the documentation and system output are different. Suppose the management cannot easily identify strengths and weaknesses within the company. In that case, it could be more difficult for a functional manager to convince upper management to make major investments in equipment and facilities needed to support a unit's technical work.

The clarity indicated for a functional structure is likely the main driver for their continued existence. However, even though this is the visible case, there are significant shortcomings here when structuring a project team. In this scenario, the functional structure greatly suffers. The functional structure does not handle moving various skills in and out of a project's needs, so embedding the project needs into this structure opens another set of requirements not well suited for this. Potentially, there could be slow response times to project requests since there will be internal politics regarding which projects get approved. People who know each other may not feel comfortable challenging the status quo. Therefore, if there is not a set of fresh eyes, it can be negative because processes and power struggles that exist may be ineffective. It can also be difficult to manage peaks and valleys in staff workloads. Project needs do not match the availability of team members in that there could be too much or too little work at any one time because all project team members come from one department. With a smaller pool of labor resources to tap into, there may be a lack of specific skills among the group, resulting in skill gaps within the project team. This can cause problems with the need to pull from another division of the company to have a labor resource that is a valuable team member and proficient in the skills needed.

Within the functional organization structure, there is a chance for overlap or duplication among projects in the same specialty area performed for more than one group within the company. Therefore, redundant work may be performed throughout the organization. For example, multiple specialty groups such as aircraft and facilities could create a system to track maintenance of aircraft and facilities. If this project were instead performed jointly, it would take less effort because maintenance tracking for either aircraft or facilities could be contained in the same developed system. This is a waste of resources in terms of labor and monetary resources. PMI (2017) summarizes the functional organization as the project manager has little to no authority. The project manager tends to be serving a project in a part-time capacity. The functional manager tends to manage the project budget. Lastly, the project manager will typically have access to part-time administrative staff to support the project.

There is also a multi-divisional organization structure that is described by PMI (2017) as replicating the functional structure but for each division of a company. Therefore, the same structure and degree of authority for the project manager exists, as described for the functional structure. The difference is that there are multiple functional organization structures within the same organization based on different geographical regions, products, processes, portfolios, programs, or customer types or however the divisions of the company are organized.

# MATRIX ORGANIZATION STRUCTURE

The matrix organization is "any organizational structure in which the project manager shares responsibility with the functional managers for assigning priorities and for directing the work of persons assigned to the project" (PMI, 2017, p. 710). Matrix organization structures involve creating project teams that can rapidly pull employees from various functional units to build a project team, just as done in a functional structure but without disrupting the structure because matrix structures were built for this (see Figure 5.3). Therefore,

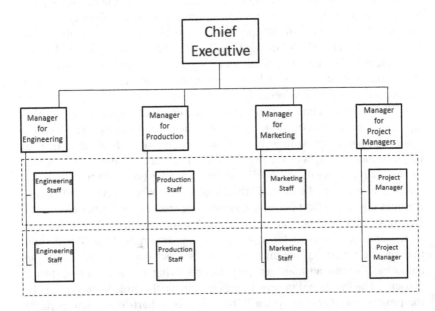

**FIGURE 5.3**  Matrix organization.

a big benefit of the matrix organization structure is that teams can be quickly assembled because there is a larger resource pool to tap into, and the organization already has employees with different expertise. Scarce expertise can be applied to different projects as needed. Therefore, the organization can justify maintaining a full-time employee with expertise because this employee can be used for small amounts of work on multiple projects. The positive aspect is that the project cost is decreased because a full-time dedicated resource for one project is not needed and the cost is shared among multiple projects. Getting buy-in from team members' functional units is easier because employees working in the functional area impacted by the project deliverable will be more likely to support the project such as providing input to help develop a better system design for a particular functional area. Another benefit is that consistent reporting systems and procedures can be used for projects of the same type since employees from the same functional areas with already set systems and procedures work on an array of projects.

The matrix organization structure can be considered a weak, balanced, or strong matrix defined by the level of authority granted to the project manager. The matrix is a combination of functional and project-oriented organizations. The stronger the matrix, the more it resembles a project-oriented organization where project managers manage the project budget and have full-time administrative staff with moderate to high authority. In a strong matrix, the project manager manages the project budget. The weaker the matrix, the more it resembles a functional organization where the project manager has low authority, part-time administrative staff, and the functional manager is also the project budget manager. The balanced matrix provides the project manager with a slightly greater degree of authority than the weak matrix but still not the full authority as found in a strong matrix. The balanced matrix also has a part-time administrative staff with a mix of the project manager and functional manager managing the budget.

The matrix organization structure also has challenges. Team members often respond to two different managers. One manager is the project manager assigning the work to be accomplished, and the other one is the employee's supervisor responsible for performance appraisals and leaves requests. If the two managers do not communicate, it could cause issues such as an employee being approved to be on leave during a critical portion of the project or the employee doing great work but not being rewarded on their performance appraisal because the work does not serve the employee's manager. In this environment, the project manager has a weakened power of control over their team because the supervisors have more ultimate power over the employees than the project manager. Therefore, the employee may be more apt to please their supervisor, who does the performance appraisal and approval of employee leave. Ultimately, the team member has more forced loyalty toward the supervisor. The project manager also must deal with multiple supervisors because any one team may be comprised of different employees, each having different supervisors. This aspect of the matrix organization structure is difficult.

Another challenge is that team members working on multiple projects might have to address competing demands for their time, such as trying to please all their managers, resulting in an employee having too high of a workload. The employee may not have a single manager looking out for their best interest but rather have one wanting the project to be completed as planned according to the outcome, schedule, and budget. Furthermore, team members may not be familiar with each other's styles and knowledge unless the same team has worked together in the past. Building trust and being comfortable with team member's work styles do not occur overnight, yet projects have a limited completion time. Team-building exercises can aid in this effort but not fully help.

There can also be a lack of focus on the project team and its goals instead of each person's assignment. Team members may have a hard time focusing on the overall good for a particular project versus what is best for their functional area. For instance, a project may be put in place to build a system that may result in greater efficiency, resulting in layoffs due to automation. A team member from the affected functional area may have a hard time keeping the project focus because their higher priority may be a concern for colleagues' jobs. Also, different team members might use multiple work processes and reporting systems which could cause internal team politics. In this situation, it would be best handled by the project manager, setting the standards so that each team member utilizes the same processes and systems. When team members are distributed geographically in different locations, the complexity of coordinating the team becomes an even more significant challenge. Typically, large and global companies tend to have matrix organization structures, including the hybrid organization structure discussed last in this chapter. The more mature a company is, the more likely it is to resemble a hybrid organizational structure.

# VIRTUAL ORGANIZATION STRUCTURE

Globalization in business is changing traditional organizational structures. The virtual organization can take on any of the structures previously described in this chapter. Virtual teams consist of "groups of people with a shared goal who fulfill their roles with little or no time spent meeting face to face" (PMI, 2017, p. 725). However, the virtual organization consists of many employees working out of their homes or at different locations. Technology allows individuals to communicate at a distance, so companies that operate virtually can operate just like any other organization. There are many benefits and challenges for virtual organizations that can range from any of the benefits and challenges affiliated with the previously mentioned structures.

A benefit not discussed previously is that costs can be lower when physical facilities are not needed. The virtual structure has become a reality due to the ease of virtual connectivity through advances in technology. Virtual offices contain distributed work teams that equate to team members being geographically dispersed and interacting at different times. This structure tends to be best for an individual within an organization that is self-managing and not in need of much supervision. It is also best for a team to communicate with a manager who serves more in a facilitator role. Organizations with off-site employees requiring expertise found in multiple locations around the globe have no choice but to conduct themselves virtually. A benefit is that a company or project team can tap into resources around the globe that best fit a project's needs. Another benefit is that there are fewer water cooler-type politics just because of the mere independence of a virtual team.

Challenges within the virtual organizational structure exist for colleagues who are not collocated due to the greater potential for miscommunication. Building trust is difficult enough with teams that are collocated versus a distributed work team. Communication can greatly suffer when there is little to no face-to-face communication. This challenge is great and can be detrimental to a company's survival if not monitored and addressed appropriately.

PMI (2017) summarizes the virtual organization in that the project manager has low to moderate authority. The project manager tends to be either operating in a full-time or part-time role as a project manager. There also is full-time or part-time project management administrative staff for the project manager and project. Lastly, there is a mix between the functional manager and the project manager that manages the project budget.

# PROJECT-ORIENTED ORGANIZATION STRUCTURE

The project-oriented organization structure is an organization where project team members report solely and directly to the project manager (PMI, 2017). There is also staff to support the project manager. Figure 5.4 depicts a typical organizational chart for the project-oriented structure.

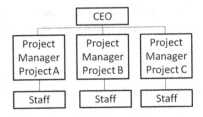

**FIGURE 5.4**   Project-oriented organization structure.

In this structure, the project manager has high to almost absolute authority. However, the next section on the hybrid organization structure can also somewhat apply to the project-oriented structure too because the project-oriented structure can also be referred to as the hybrid or composite organization structure according to PMI (2017), but the hybrid is also listed as a separate category too. The project manager role is a full-time role. In this organizational structure, the project manager has control over the project budget. There is often full-time administrative staff dedicated to supporting the project manager and the project. The project-oriented structure is also a combination of the other structures discussed in this chapter, but the characteristics of this structure are most similar to the strong matrix.

# HYBRID ORGANIZATION STRUCTURE

A hybrid organization is a mix of other types of structures all within one structure (PMI, 2017). It has units resembling functional, matrix, project-oriented, and virtual organization structures, as displayed in Figure 5.5. This organizational structure was developed to meet or customize the structure to the specific needs of a company. The main benefit that exists is the same as in the functional organization structure because of standardization throughout the organization regarding processes and reporting systems. Challenges within this organizational structure tend to be in increased conflict within and between projects because of duplication of effort and overlapping authority.

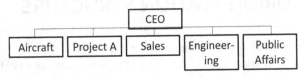

**FIGURE 5.5**   Hybrid organization structure.

Many managers have issues with performance appraisals and leave requests residing with one manager and where one or more additional managers are responsible for assigning work tasks. Typically, large and global companies tend to have matrix organization structures, including a hybrid organization structure. The more mature a company tends to be, the more likely it is to have a matrix or hybrid organization structure.

# SUMMARY

For many organizations, power and political conflicts occur due to labor, monetary, facility (space), and equipment resources. Every organizational culture is different, but every successful project needs support from the organization so that the organizational structure is supportive and not restrictive. Different organizational structures serve different needs, so companies will select the structure or combination of structures that best fits their needs. Depending on the structure in place, different power struggles and politics will occur. However, even though different structures have different power struggles and politics, no structure is without these issues. Different organizational structures were addressed, including functional, matrix, virtual, project-oriented, and hybrid. Benefits and challenges exist within every structure; companies organize themselves based on which organization fits their company the best.

# REFERENCES

Maddalena, V. 2012. A primer on project management. *Leadership in Health Services* 25(2):80–89.
PMI (Project Management Institute). 2017. *A Guide to the Project Management Body of Knowledge*, 6th ed. Newtown Square, PA: PMI.
Richardson, G. L., and Jackson, B. 2019. *Project Management Theory and Practice*, 3rd ed. New York: CRC Press, Taylor & Francis Group.

# Defining the Project Scope

# 6

## INTRODUCTION

PMI (2017) defines scope as an activity to identify "the sum of the products, services, and results to be provided as a project" (p. 722). This description contains elements related to both an output goal and the project work related to that goal. The process of creating a valid project scope is often mismanaged, and this deficiency cascades throughout the subsequent life cycle. Think of this activity as the foundation of a house. It is very hard to have a stable house without an appropriate foundation. Failure to execute this definitional process properly leaves the project in a very unstable position for technical and management success. This chapter describes three basic steps in creating a project scope from which all key parties will review and approve.

The three sequential steps described here trace the project requirements through increasing levels of definition. A formal name for this is *Progressive Elaboration*. The process starts with a somewhat vague vision and moves through increasingly detailed steps to a more technical level of detail required to execute the task. The final step involves approving the requirements and integrating them with available resources. *The important idea here is to understand that future topics such as schedule and cost should never be formalized until a clear picture of the required work is defined.*

The key planning stage terms involved in this process are as follows:

- A Charter is a formal document issued by the project owner who approves the project start.
- A Business Case is a document that describes the vision for the project and the estimated cost/benefit.
- Scope Management Plan is a document that adds technical and management definitions to the requirements.
- Work Packages (WPs) and the Work Breakdown Structure (WBS) are key working-level requirements documents that provide technical work direction for execution and control.

DOI: 10.1201/9781003218982-6

Decisions made through each of these layers should be formally documented as future results will be compared to this documentation set.

# GETTING STARTED

Step one involves establishing the Business Case, outlining why the project has merit and providing a high-level estimate of the required resources. After a screening process, senior management may decide to support moving this initiative to the next level. The formal document to show this approval is called the project Charter. A good management rule to follow is never to commence a project without a formal Charter. This document formalizes and validates management support and approval for the effort. It also defines key output goals for the project, as well as other constraints and management guidance. More details about this process are available in Chapter 5 in Carstens and Richardson (2020).

## Use of Templates

In many areas of the project, there is a need to document various views of the process. This can be time-consuming and is often sluffed as not productive. These documents all have a communication role that should not be ignored, and one way to make the process more palatable is to use pre-formatted templates. These can be fetched from other similar projects or extracted from numerous external sources. More details on these sources will be shown in the next chapter, but realize that a preformatted template with appropriate "boiler-plate" and structure saves time and improves the project plan view.

## Business Case

Early in the development of a project vision, it is common to produce a document called a Business Case. The role of this document is to translate the initial vague vision into a high-level analysis of the tangible and intangible value of the initiative. This document provides necessary financial estimates, broad risk assessment, and goal justification to support the approval for the expenditure of resources in competition with other project initiatives that are also being considered. In all project situations, the availability of resources will constrain the level of project activity that the enterprise can support.

A Business Case is comprised of the following sections (Richardson and Jackson, 2019, p. 180):

- Project objective
- Problem/opportunity statement
- Potential solution strategy
- Organizational goal fit
- Strategic goal fit
- Key assumptions
- Competitive analysis
- Benefits (tangible and intangible)
- Cost estimates (development life cycle and possibly production maintenance)
- Competition
- Recommendation

The appropriate audience participating in developing a Business Case can be varied and depends upon the perceived scope of the effort. If the proposal is initiated from the bottom of the organization, the next step would be to seek a higher-level manager who will support the idea.

# DEFINING REQUIREMENTS

The formal requirements definition process starts from the guidance found in the Business Case and the Project Charter. Some organizations require an initial requirement gathering to verify the potential value, while others outline the vision at a high level. Regardless of the starting point, realize that much more refinement is needed before the approved project scope is determined.

Here is a slightly modified actual scope-related example to show how *not* to do this. The story starts when a senior manager announces a promotion to project manager. The promotion is for the approved Project XYZ for which an earlier Business Case has been produced. In this document, the proposed schedule is one year and a budget of one million dollars. The actual requirements have not been well defined at this point. With this scenario, here is the key message to remember: if you are the target promotion, you must explain to the project owner that this initiative does not yet have sufficient scope definition to verify the schedule and budget goals. Then, go on to explain why this is necessary. State that you will do everything in your power to meet these goals and will come back at some specified time with details of the analysis process.

If you don't take this stand now, you are likely to accept a set of goals that have no basis of being achievable. In other words, your first project will be a failure. As a memory device, call this scenario *Project Titanic*. Believe it or not, this is an all too frequent occurrence.

*The Planning Rule: All project plans will be based on a formally approved set of requirements, and there will be direct links from the initial request into an approved deliverables plan.*

The next project management step involves translating the initial vague language into a physical work definition. Second, the requirements process must seek agreement between the project owner, other stakeholders, future users, and the technical team assigned to produce the outcome. This process requires more specific input than previously described in the Project Charter or Business Case. Given many different perspectives of the project, consensus-building is a key activity at this stage. As this process unfolds, a document called the *Scope Management Plan* (*SMP*) should be created to contain a written description of the ongoing specifications defined.

Here is a memorable example to highlight why formal analysis and documentation are required for this step. Suppose the hypothetical Charter stated that the desired product was supposed to

*Leap tall buildings with a single bound.*

We will also say that the project team knows how to build this sort of product (i.e., building jumpers). It should not be hard to see that goals such as this are not adequate for technical or managerial purposes. Think of this next step as clarifying that knowledge gap from early vision to later work required. As new specifications for this goal are received, they will be documented in the SMP. The list below provides some sample SMP-type documentation of technical parameters:

- How tall is the building?
- Does it need to be successful every time?
- How much training would be needed to operate the device?
- What cost market is the device supposed to fit into?
- How soon is the product needed?
- How will the project be managed (i.e., status reporting, changes, etc.)?
- Describe the resources required and locations involved.

Note that questions of this type are aimed to provide more succinct guidance for the technical team as they attempt to define further how to produce the proper output device. For each of the sample questions, the SMP should strive to define quantitative goal values to be used in future planning and tracking stages. Customers often feel that simply stating the high-level goal is sufficient, but nothing is further from the truth. Details such as the list above are necessary for the project team to resolve as they jointly develop the project work specifications. Failure to define the required work through questions of this type will almost surely create future confusion regarding the resultant product. The process of collecting requirements is fraught with communication issues. Traditionally, this process was accomplished by interviewing different groups and then compiling a summary requirements document. This method tends to be ineffective for numerous reasons—time to accomplish the collection being the most noticeable. This serial approach method essentially focuses on a one-way sequential input view that omits negotiation across various parties. Unless a consensus is obtained, this format can lead to another step along the Project Titanic trail. A more modern method of executing this process is to hold a joint session of key players whose role is to lay out more details regarding how the project will be viewed. This is somewhat similar to an expanded Business Case with some project structure added. The focus here needs to remain on *what* is to be done. Once a requirement definition is approved, the next step in this process is to divide the defined work or product deliverable units. From this level, the process moves downward into increasingly lower-level segments. This is another example of progressive elaboration. This joint review process will likely generate controversy among the group, but a good leader can usually work through these issues reasonably fast. At the end of this cycle, a full description of the project work components should be observable. The project vision is still fairly high level at this stage, but now a macro-level work structure is visible. As a real case example, a $2-billion oil drilling project was work-defined in three days using key stakeholders. Output specifications were not quantified at this point, but the broad scope of the effort was laid out. We'll see more of the elaboration mechanics below. At this stage, there is general agreement regarding scope of the effort.

## Scope Management Plan

As more data is collected regarding project scope, it needs to be captured in the SMP. Throughout this stage, the ongoing goal is to define the effort in quantitative terms as much as possible. There are four key output areas for a project that need a clear definition:

- Scope—what
- Time—when
- Cost—how much
- Risk—identifying factors that have the potential to decrease the success of the project

Also, other factors can be added to this type of decision data. This could be any factor that might affect the project in producing the defined output. As an example, the project will be conducted in City X. The following is a sample list of definition items needed for scope definition:

Project and product objectives
Product characteristics
Project boundaries
Project deliverables
Project constraints
Project assumptions
Preliminary cost estimate
Approval process

One of the definitional requirement perspectives that often goes overlooked is what is called the *abilities* list, which is summarized below (Carstens and Richardson, p.51):

1. Traceability—documenting specific configuration specifications
2. Affordability—matching desired function to cost
3. Feasibility—any adverse item that can get in the way (i.e., technical, organizational, political, economic, resources, etc.)
4. Usability—breadth of value for the outcome
5. Producibility—how hard is the item to make
6. Maintainability—how much maintenance is required
7. Simplicity—a measure of technology complexity
8. Operability—how much user training is required
9. Sustainability—how long should the item maintain operational value

Always review this list to verify that the defined requirements match the planned goal. Guiding statements such as this should be defined early in the process.

# PROJECT SCOPE DEFINITION

Beyond documenting deliverable requirements, the scope definition also deals with project boundaries that specify what should and should not be included. Other items to be included in the scope definition are as follows:

- Deliverables—products, services, documentation, and other management artifacts.

- Acceptance process—how will the outputs be evaluated.
- Constraints—these can be any defined limits to be observed by the project.
- Assumptions—it will be necessary to make assumptions in the planning process. These should be documents and reviewed throughout the life cycle.

## WPs and the WBS

Now that the high-level view of the project has been discussed with key stakeholders, the next step in the process involves creating sufficient work definition to facilitate further planning activities related to work specifics, scheduling, budget, and control processes. Two critical tools are most useful in the scope development process, *Work Breakdown Structure* and *Work Package*. Brief definitions of these are as follows:

- The WBS is a popular method for graphically describing the project scope and work architecture. This is a graphical view that identifies project structural elements and work groupings. A supporting dictionary is also recommended.
- A Work Package is a management-sized element of work that can be reasonably estimated and measured.

A useful mental image of a WP is to view it as a box-shaped project building block with dimensions of time, cost, and functionality. If you were a chemist, this view would be a molecule view of the total compound. In this analogy, as the scope increases, the box size also increases for schedule and cost. Another somewhat more complex idea is that it is possible to expand a box (scope) and change the project characteristics. Also, recognize that by stacking small work boxes into time periods, both schedule and cost can be estimated. Recognize that there is a technical sequence for these boxes to be executed. The box metaphor is an important conceptual step in understanding the planning process. From a pure theory viewpoint, if the dimensions of each work box could be accurately defined, a quantitative measure of the project scope is determined. Also, if these boxes could then be stacked into some appropriate work sequence, a schedule is produced. More explanation is needed, but this is a good starting point for planning the project. From this critical scope information, it is possible to not only plan but also track the status of the effort.

This theory represents a valid design model for all project methodologies. However, it is important to recognize that these estimates have potentially high error rates for a myriad of reasons. Nevertheless, *this is the base model and represents the starting point for project management.*

Recognize that some project types can define work estimates at a low level of detail much better than others. For example, a home builder can do an excellent job of defining work units and do a reasonable job of estimating the associated work effort. However, pursuing a project to produce a new cancer drug does not fit this model well. Inability to accurately plan a project scope hampers future project control and endpoint forecasting ability.

One of the common dilemmas for a project manager is to deal with is the sponsor who wants to know up front the schedule and cost for the project. Having any hope of doing this accurately is based on the appropriate scope definition. Other variables are tied to this, and they can also be difficult to estimate. Chapters 7, 8, and 9 will shed more light on this evolving planning process.

# CONSTRUCTING THE PROJECT SCOPE

At this juncture, we are at the chicken and egg stage. Do we define the work packages first and then integrate the pieces, or do we envision the effort at a high level and then break it down into small and more manageable units? There is no singular correct answer to this, but our bias is to start at the top level and break down components or stages into increasing levels of detail. The process is called *decomposition*. If previous experience has unveiled what such a project entails, building this structure from that historical background is possible. With little or no background, the structuring process likely involves a little of both directions until the overall structure seems coherent. Regardless of the development method, the end goal is the same—one clear structure showing how all pieces fit. We will illustrate the top-down approach here to create a project WBS.

# WORK BREAKDOWN STRUCTURE

The WBS is one of the more familiar acronyms in project management. Recognize that the term is used in different ways, but the description here is a "hierarchical tree structure" for outlining the defined project scope and related work. The higher levels represent major aggregations of the project, while the lowest level represents actual work activities that collectively are needed to produce the deliverable scope. There are many technical references available

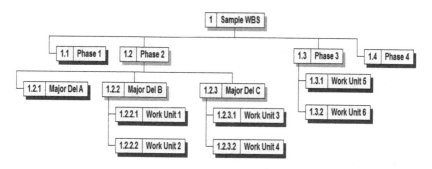

**FIGURE 6.1**   Structuring the project deliverables.

to pursue more advanced mechanics for developing a WBS. However, in reality, common sense may be the best rule. If the design structure fits the way you see the project, then it serves the right purpose. This section illustrates a few mechanical examples and rules to help you better understand this process.

Figure 6.1 shows an abstract WBS example showing how project phases and major work deliverables can be structured. The list below contains a few WBS design notes:

- The top box represents the total project (scope).
- Phases and major deliverables are used to segment key subsets of the project.
- Lower-level boxes represent actual work tasks at the bottom of the structure.
- Planning boxes may be used to show future work that is not yet approved or fully planned.
- Summary boxes can be used to accumulate groups of lower-level work segments (the term Major Deliverables is used in the structuring example).

Before moving from this definitional level, a few important related WBS concept rules also need to be understood. Summarized below are some key ideas that should be followed:

1. If a work item is not defined in the WBS, it is not in the project (i.e., only do what has been approved).
2. Once the project owner agrees with the WBS, the project manager should sign a formal acceptance letter. This essentially becomes a contract between the two parties.
3. An approved WBS defining the approved scope is *baselined* (essentially frozen), which means only formally approved changes will be added moving forward.

4. Any changes to the project scope must be formally approved and added to the official WBS for tracking purposes. Approved changes often add extra work, plus add additional time and budget to the plan (but this is seldom recognized).

5. The project schedule and budget will be derived from this structure (Chapters 7 and 8), and the future status will be tracked and compared to this baseline.

# WORK PACKAGES

As shown in Figure 6.1, the boxes labeled as work units represent the actual work required to complete the effort. To clean up our vocabulary, we now need to define these work units as *WPs*. Keep in mind the box metaphor shown earlier. A working definition of a WP is as follows:

*A moderately sized item of work related to satisfying some element of the project scope.*

While there is no fixed standard as to the size of a WP, a general rule of thumb is 80 hours of work and two weeks of effort. Using larger-size packages can compromise the ability to measure status. Here is a memory example for this point:

*Imagine a WP that has a duration of one year. When the estimated due date arrives, the team has claimed that the status of the WP is 90% complete. Later, as the WP is completed, the amount of time and resources consumed required another 90%. To avoid this 90/90 syndrome, it is better to have small units to track, decreasing tracking measurement errors.*

A review of this definition shows that the WP role is multidimensional. It initially serves to define the required work, but it also has a key role in further planning steps and supports status monitoring, as we shall see in various other areas. To serve this broad role, the WP definition includes a technical description of the work involved and data related to factors such as work effort, duration, resources required (skill and quantity), estimated cost, and technical sequence. A general theory of project management says if you cannot define the required effort to execute the project, you have nothing tangible to monitor and control future direction.

Fundamentally, the role of project management is to deliver what has been defined. In traditional terms, this means the defined *scope, schedule (time),* and *budget.*

*A sidenote.* The project repository is one of the most valuable artifacts in project management. Its role is to formally capture decisions made into a formal storage location where current and future projects can extract data. This is a practice often not followed but certainly ranks as an excellent operational goal. This collection of data can be invaluable in various phases of the project and also of value to other similar projects. Project teams often spend too much time recreating the same data from one project to the next. Recognize that many decision elements are reusable, and it would save considerable time if the data were readily available for reuse. Similarly, various pre-defined templates can be reused along with this data (i.e., proposals, project plans, WBS, status formats, etc.). *Recommendation—Spend some time defining how and where you want to store your collected project artifacts.*

Let us now show how the WBS and WP fit together to describe the project scope and work structure.

## BIKE EXAMPLE

Using the project design represented by the WBS and WP, we can illustrate how these can be used to produce a scope overview for constructing a bike. Figure 6.2 contains a sample WBS scope overview of the bike project.

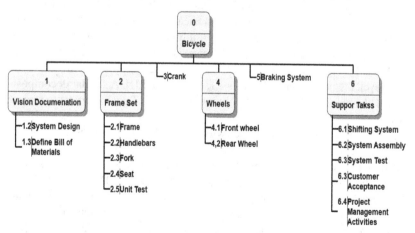

**FIGURE 6.2**   Bike WBS.

Note that this describes how the project is envisioned. In this example, the structure is primarily based on the bike components—a product structure. An alternative design could have defined this as a process layout. Also, note the left and right sides of the structure describe supporting work (i.e., design and support elements). These are added to show the full project scope. This is the essence of a WBS in a practical view. More theory and examples of this approach can be found in Chapter 8 of Carstens and Richardson (2020) and Chapter 12 of Richardson and Jackson (2019), as well as other sources.

Each of the high-level "box" segments shown is considered to be a phase or major activity area. Items grouped below the major segments are WPs that fit our earlier description. In this design, we assume that none of the WPs are large enough to require further decomposing for estimating or status tracking purposes. It may take a little practice to get the hang of this idea, but it is certainly worth doing to help envision the full project view. Once such a view is created, future work efforts are often described using the WBS numbering scheme.

# WBS NUMBERING SCHEME

To add a little more reality to this approach, let us look at a more complex WBS, as shown in Figure 6.3.

In this view, only project phases 3 and 4 are highlighted. Phase 1 may be completed and shown here only for historical reference. Phase "n" is a future undefined effort that we might show to imply that a future phase is envisioned but not planned as yet. The top level is used to layout the bigger picture. Note the top box is labeled with "1." As the structure decomposes downward to the next level, the numbering scheme is 1.1, 1.2, 1.3, and 1.4. A similar decomposition of the numbering scheme continues downward. Take a moment to review how the lower levels inherit their number. This is an important concept in the definition, and these number codes will be useful in many places later.

As shown here, a deliverable can be anything that is a focus item. Major subsystems are a good example. Once again, note how the numbering system shows the layers. As an understanding test for this idea, Deliverable 1.2.2 is the second deliverable under 1.2 (phase 2). This coding scheme can be used to identify where in the structure a particular box fits, and the graphical view will provide a good visual overview of the total project. WPs appear at the bottom level. Some theorists might quibble with defining box 1.2.2.2.1 as a WP since it does not occur at the lowest level. However, if it makes sense to the team, it is a minor issue. A more proper term for 1.2.1.1.1 is a Control Account to indicate a group of WPs used for management purposes. The key design concern here

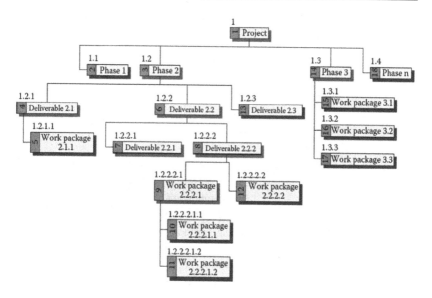

**FIGURE 6.3**   WBS numbering example.

is to say that the WBS should mirror the way the project team plans to manage the project. Understand that from this important base point, we continue the elaboration process to focus on work estimating details regarding the WBS.
*Note: One key editorial item for this section. The top level of the WBS can be labeled with either a "0" or a "1" code. If you choose the "0" option, the next level boxes would be coded as 1,2,3,4. The levels below that would pick up the same numbering logic as before.*

## Standard WBS Templates

It is common for organizations doing similar projects to create standard artifacts (i.e., Business Case, Charters, WBS, etc.). This process helps with future estimating using historical data, and it also helps communicate common status across similar projects. Table 6.1 shows a sample WBS for software development. For organizations that perform similar projects, standard templates save significant structuring time and this helps make cross-comparisons possible.

Note that the top box WBS code for this design is labeled "0" in the template (instead of "1" as in the previous example). It would be a good reader exercise to draw the hierarchical schematic WBS from this list of numerical codes.

**TABLE 6.1**   Software Development Template

| WBS | TASK NAME |
|-----|-----------|
| 0 | Product name |
| 1 | Planning |
| 1.1 | Initiation |
| 1.2 | Meetings |
| 1.3 | Administration |
| 1.3.1 | Standards |
| 1.3.2 | Program office activities |
| 2 | Define product requirements |
| 2.1 | Define requirements |
| 2.2 | User documentation |
| 2.3 | Training program |
| 2.4 | Hardware |
| 3 | Detail design |
| 4 | System construction |
| 5 | Integration and test |
| 6 | Project management |

# SUMMARY

This chapter has described an abbreviated summary of steps necessary to create a project scope structure. From this foundation definition, the rest of the project planning and control elements will be attached. Remember, the WBS is considered the fundamental foundation artifact for the project.

# REFERENCES

Carstens, D. S., and Richardson, G. 2020. *Project Management Tools and Techniques.* 2nd ed., Boca Raton, FL: CRC Press.

PMI (Project Management Institute). 2017. *A Guide to the Project Management Body of Knowledge,* 6th ed. Newtown Square, PA: PMI.

Richardson, G. L., and Jackson, B. M. 2019. *Project Management Theory and Practice,* 3rd ed. New York: CRC Press.

# Building the Project Schedule

# 7

## INTRODUCTION

This chapter outlines the key approaches for two basic plan development activities: estimating work and constructing the project's initial raw schedule. Microsoft Project formats are used as examples; however, they mirror a manual view. The approach outlined here is designed to maintain integrity of the approved scope definition from the previous chapter, where Work Packages (WPs) and the Work Breakdown Structure (WBS) defined the approved scope.

There are two major steps in this schedule-oriented segment. The first step involves translating the approved scope definition artifacts, WBS and WP, described earlier, into quantitative parameters required to produce a first cut schedule. Step two involves the mechanics related to the parameters needed to create a schedule. Underlying this process is the requirement that the plan derived matches the approved scope definition, and this ensures future status tracking integrity from that base point. In other words, the plan is not just a pipe dream of wanting certain tasks to be done at the desired time, but one that links directly to the scope definition artifacts. This linkage process is designed to make the resulting plan integrate with the approved project structure (WBS), task estimates, defined sequencing, and finally use this to assess the adequacy of available resources.

## ESTIMATING PROJECT REQUIREMENTS

The process of accurate project outcome estimating is complex for many reasons. The most obvious of these involves translating a defined goal requirement into equivalent work time. Beyond a work sizing activity, there is also the need to estimate the skill and productivity of the assigned resources, even though a specific worker is unknown at this point. It is

DOI: 10.1201/9781003218982-7

important to recognize that estimating errors will create schedule calculation errors in the resulting plan.

One of the first confusing issues with estimating is the vocabulary of an estimate. When one asks, "how long will this take," we have our first vocabulary problem. Do you mean days, hours, or calendar (recognize that the team may not work every day)? Managers are typically interested in calendar estimates, while supervisors may be more inclined toward hours or workdays. Project plans focus on calendar dates, but the estimating process must start at a much lower level and build upward. The estimating examples in this section will focus on duration in workdays.

## Estimating Components

Estimating involves translating the approved requirements into resource levels, skills, risk, and other factors. The normal goal is to quantify worker time, skill, material needs, and other related resource costs. Individuals who can accurately do this are industrial artists. Another further complication occurs with the use of both internal and contracted resources to execute various tasks. It is important to segregate the estimates by type. A third category of resources to estimate deals with items such as travel, supplies, and support equipment. Beyond this, there are also the infamous overhead charges to the project. This chapter will focus on schedule parameters, and the next chapter will use this base data to move the focus toward the cost side of the project estimate. One final key point regarding how estimates are structured is to point out that WPs will be the target for estimates, and this same target source will continue into the budget side.

The chain of events surrounding the schedule estimating activity is as follows:

1. Customer approved scope based on WPs and WBS.
2. Rigorous estimating of WPs that represent the total scope of the project.
3. Construction of a work plan schedule from WPs and WBS.
4. Evaluate human resource availability to meet the requirements.
5. Balance the plan to match available resources. Omission of this step means that the plan is a wish list rather than a viable estimate.

One simple way to illustrate some basic estimating issues is to use an easily envisioned work activity. Let's assume that creating a hole in the ground is a WP for a larger project. The hole specification is 10 × 10 × 10 cubic feet. That is a very visible physical work task. What is your estimate for digging this hole? What are some questions that might help you answer this? Let us first say that this hole requires moving 1,000 cubic feet of dirt. A work hour total could be computed if we could estimate cubic feet per hour. Assume that this raw work estimate is 48 hours. Does this mean six eight-hour days on the schedule? How long would it take using two resources? What is the work schedule for the resources? Does this estimate include rest times, etc.? What if a mechanical aid were used instead of manual shovels? With all of these potentially confusing variables, what value do you assign to this task? We could carry this simple example along much longer, but it is pretty obvious that even this basic example has many related questions and assumptions that affect the derived answer. We also see that work calendars are important to create plan dates, and we see that the type of technology used can impact the answer. What is often ignored is the potential for some external probabilistic environmental impact, such as weather. Recognize that having a skilled estimator is certainly an important support service for the project and also recognize that one should still expect to find variations in these estimates as the project is executed.

## Estimating Techniques

There are various ways to estimate project work, and the correct answer for this involves the type of project, historical experience with that type, resource availability, and many other factors. Four of the most common estimating methods are described here. More details on other methods can be found in Chapter 8 of Carstens and Richardson (2020). The sample methods described are as follows:

> *Expert Judgment.* This technique is based on the skill and judgment of the estimator. This is most accurate when there is a repetition of project types (i.e., houses, products, etc.).
>
> *Analogous (Top-down).* This is similar to the expert method, except it is usually done at a high level and uses data from similar projects. For example, if there is historical data and experience with office buildings, an estimate for a similar five-story building could use previous data as a guide because there would be common elements usable in the new estimate. This generally would lead to a high-level estimate that would still need to be broken down into smaller WP units.
>
> *Bottom-up (WP Level).* This method starts at the bottom of the WBS, and WP-level estimates are made using multiple techniques. That collection would then be summarized upward to generate the top project estimate. Literature suggests that this should yield the most accurate estimate.

*Heuristic.* This is called a *Rule of Thumb* method. Suppose that a local roofer bids on a project to redo a roof. He has done several roofs in the area and the related material is similar. Only the roof sizes vary. Based on simple parameters developed from previous experience, the roofer has found that a measurement of the roof can be used with a defined factor per square foot to estimate the cost. A similar approach can be used to estimate home construction time and cost, using similar metrics such as house size. Auto service shops offer another example. When the job is to replace an alternator, the estimate given will be taken from a book of parameters derived from historical data.

Once estimates have been made to the WP level, a checklist such as the one below should be used to evaluate the results:

1. Do you have sufficient requirements definition for the project, including management and various support areas?
2. Do you have a fully decomposed WBS (failure to have this opens up potential gaps in work definition)?
3. Do you have historical information on similar projects?
4. Have you identified all scope and resource elements for the project (i.e., defined labor skills, materials, supplies, equipment, etc.)?
5. Do you have justifiable reasons for selecting your estimating methods, models, guides, or commercial software?
6. Have you considered risk issues in your WPs and overall plan?
7. Do your estimates cover all tasks in the WBS? If not, recognize the potential errors resulting from these gaps.
8. What level of accuracy is required for the estimate?

Be aware that there is much more to estimating work than is covered here. Large companies have full-time employees doing this type of work, and it is a professional job title. Regardless of how the project estimates are derived, the key items outlined here are important factors to consider. The level of accuracy to expect varies based on project type, but expect variances during execution.

## PLAN GENERATION PARAMETERS

A warning was made earlier about not accepting an early target schedule for a project without detailed scope and estimating data. We are moving in the right direction now, but the question is, what data and estimates are needed to construct the initial plan? The following WP-level data are needed:

- WP Name
- WP duration estimate (based on allocated resources)
- WP sequencing (define the WP order of execution)
- Work calendar for project and worker (i.e., holidays, weekends, vacations, etc.)
- WBS code—this is optional but useful for task organization

Before diving into the formal plan calculation, let's quickly look at the most used format for project plans. In the early 1900s, Henry Gantt created a graphical planning document to show visual schedules. That format is called a Gantt chart, and it remains the most used planning tool format till today. Figure 7.1 is a simple example of this chart format.

Most would agree that it takes zero time to learn how to interpret this view, which is the key to its long-term success. The format shown is not exactly the original format as there are two modern extensions shown in this view. First, this chart is created by Microsoft Project, which is the most used project management software in use today (obviously, not available at Gantt's invention time). Second, the bars now have intertask linkages to show technical sequencing. There are other similar software tools on the market today that will generate a graphical Gantt chart. Regardless of the underlying mechanics used to generate a schedule, it will have to be formatted like a Gantt chart to be acceptable. Realize that a Gantt chart need not have underlying integrity to be produced. The format is fine, but it should be created as a result of building the work elements as described here.

Translation of work hours and days into elapsed calendar times is time-consuming and best left to a software tool. If one plans on doing much with project management, getting familiar with a software tool will improve the ability to easily manipulate schedule calculations. In order for a plan to have integrity, it must have good time estimates, defined task linkages,

| | ❶ | WBS | Task Name | Duration | Predecessors | Jul 7, '13 | Jul 14, '13 | Jul 21, '13 | Jul 28, '13 | A |
|---|---|---|---|---|---|---|---|---|---|---|
| 1 | | 1 | Total Project | 20 days | | | | | | 8/1 |
| 2 | | 1.1 | A | 5 days | | 7/11 | | | | |
| 3 | | 1.2 | B | 5 days | 2 | | 7/18 | | | |
| 4 | | 1.3 | C | 5 days | 3 | | | 7/25 | | |
| 5 | | 1.4 | D | 5 days | 4 | | | | 8/1 | |

**FIGURE 7.1**  Sample Gantt chart.

evaluation of available resources, and a work calendar. Microsoft Project is the industry leader in this role, but it is not what one would call a trivial tool to use. This text does not explain the keystrokes necessary to produce the figures shown due to space limits here and the ready external availability of good instruction tools for this product. It is possible to use MS Project either in elementary mode or exercise its full function set. The latter approach requires a more significant time commitment. The examples show that Microsoft Project produces a Gantt format to satisfy user bias, but the bars produced have calculation integrity based on time estimates and defined task linkages.

The bars shown in Figure 7.1 represent WPs, and the dates attached to each bar show the calculated completion schedule. From a communication viewpoint, this is a very clear picture of the plan, which explains its popularity. However, from a sophistication viewpoint, one could argue that this figure is misleading in many ways. For example, will this project likely finish exactly on 8/1? Probably not, and it would be more accurate to recognize this in some probabilistic format. Nevertheless, the plan as shown represents the project roadmap goal to be tracked, even if the ongoing results are not likely to be exactly met. We now need to move on to describe some additional steps to completing an approved plan. We'll call this view the first cut version.

# PLAN MECHANICS

To make this discussion a little more real-world, we need to increase the size of the project and move away from an overly simplistic view. Figure 7.2 shows a flattened WBS with 33 named items.

This version moves the scope to a level that makes manual processing infeasible and better illustrates an appropriate role for software in the process. This example also illustrates the data elements required to construct the plan mechanically. The following list contains a few mechanical points that are not obvious at first glance:

1. Note that only low-level items will have duration and sequence information. Duration will be calculated based on sequence (precedence) information provided for each WP. For example, WBS 1.3 duration is calculated as the sequenced calculation sum of the three tasks below. Tasks may not always be in sequence. For instance, if they were parallel, the duration of this same task cluster would be seven days.

| A | B | C | D | E | F |
|---|---|---|---|---|---|
| | 2 | 1.1 | Hardware Selection | | |
| | 3 | 1.1.1 | Determine technical hardware specifications | 4 days | |
| | 4 | 1.1.2 | Selection of hardware vendor | 1 day | 3 |
| | 5 | 1.1.3 | Test hardware that is selected | 8 days | 4 |
| | 6 | 1.1.4 | Procurement of hardware | 10 days | 5 |
| | 7 | 1.2 | Software Selection | | |
| | 8 | 1.2.1 | Licensing negotiations with software vendors | 5 days | |
| | 9 | 1.2.2 | Test software from each vendor | 8 days | 8 |
| | 10 | 1.2.3 | Procurement of software | 10 days | 9 |
| | 11 | 1.3 | Integration of hardware and Software | | |
| | 12 | 1.3.1 | Implement new security model | 5 days | 10,6 |
| | 13 | 1.3.2 | Customization for each organization | 7 days | 12 |
| | 14 | 1.3.3 | Testing of the integrated desktop | 10 days | 13 |
| | 15 | 1.4 | Verifying existing infrastructure | | |
| | 16 | 1.4.1 | Test server capacity | 5 days | 14 |
| | 17 | 1.4.2 | Test the network hardware and software | 5 days | 16 |
| | 18 | 1.4.3 | Procure necessary infrastructure upgrades | 10 days | 17 |
| | 19 | 1.5 | Training for support team | 20 days | 14 |
| | 20 | 1.6 | Create documentation materials | | |
| | 21 | 1.6.1 | Technical architecture documentation | 8 days | 14 |
| | 22 | 1.6.2 | User manuals | 20 days | 21 |
| | 23 | 1.6.3 | Review documentation materials | 2 days | 22 |
| | 24 | 1.7 | Conduct user training | 12 days | 20 |
| | 25 | 1.8 | Deployment by organizational area | | |
| | 26 | 1.8.1 | Marketing deployment | 8 days | 24 |
| | 27 | 1.8.2 | Engineering deployment | 8 days | 26 |
| | 28 | 1.8.3 | Finance deployment | 8 days | 27 |
| | 29 | 1.8.4 | Executives deployment | 8 days | 28 |
| | 30 | 1.8.5 | Legal deployment | 8 days | 29 |
| | 31 | 1.8.6 | IT deployment | 8 days | 30 |
| | 32 | 1.8.7 | Administration deployment | 8 days | 31 |
| | 33 | 1.9 | Project complete | 0 days | 32 |

**FIGURE 7.2**  Expanded example WBS.

2. Note that the task sequence (predecessor) is coded by line number. For example, ID 1.3.2 has a predecessor number of 13, meaning it is sequenced after ID 1.3.1. This coding scheme is simply to save space in defining the task sequence.

3. Other sequence codes are available (i.e., parallel start); for instance, if you wanted task 34 to start simultaneously as task 23, the predecessor code for ID 24 would be 23SS (Start Start signifies that both tasks start at the same time). The role of the various sequencing codes is to drive the scheduling logic.

4. A zero duration for the last task is treated as a *milestone*. These would be tagged as special points in the project, such as a design review or customer approval. They are not viewed as a specific work task so much as a review point.

As the size of the WBS list grows, so does the complexity of coding, vocabulary, and scheduling mechanics. The three sample line items below will illustrate how the software knows how to handle various WBS entities. Sequence coding has to be manually defined in Microsoft Project. As a test of understanding, match the listing below to the WBS shown in Figure 7.2.

| ID | WBS | TYPE | EXPLANATION |
|----|-----|------|-------------|
| 1 | 1 | Project | A title defining the project name |
| 2 | 1.1 | Major Phase | A major phase of the project |
| 3 | 1.1.1 | WP | Work Package with duration and sequence |

As one can see from this example, projects of reasonable size do not lend themselves to manual construction methods. Doing more than creating a crude Gantt format requires software support. Also, recognize that project dynamics create constant change to plan views which further justifies software help (i.e., changes would be done with simple keystrokes). The technique described in this example is expandable and can be used to create much larger WBS size views. More specifics on the mechanical construction process using Microsoft Project can be found in Carstens and Richardson (2020), and various YouTube video tutorials describe similar keystrokes in great detail.

# FIRST CUT PLAN

Figure 7.3 uses software logic to translate the WBS data from Figure 7.2 into a first cut project plan.

The first thing to note in reviewing the software-generated output is how the plan result obeys the WP specifications predecessor codes in calculating the calendar dates. This illustrates the tight linkage between the scope definition and schedule. Also, note the top project title and phases (task summaries) are marked with a heavy black Gantt-type bar. Summary bars are simply collections of tasks. WP bars are coded as hashed black bars indicating that these tasks are on the *critical path**, which is another complex calculation value derived from the software. The critical path represents all tasks that are

---

* *Note in default mode, the actual software-generated critical path would be marked in red, non-critical tasks marked in blue, and major groups shown in black. This has been modified here for a B/W format.*

| ID | WBS | Name | Duration | Predecessors | Gantt |
|---|---|---|---|---|---|
| 1 | 1 | Desktop Refresh Project | 121 days | | 9/14 |
| 2 | 1.1 | Hardware Selection | 23 days | | 4/29 |
| 3 | 1.1.1 | Determine technical hardw | 4 days | | 3/30 - 4/2 |
| 4 | 1.1.2 | Selection of hardware vend | 1 day | 3 | 4/5 4/5 |
| 5 | 1.1.3 | Test hardware that is select | 8 days | 4 | 4/6 - 4/15 |
| 6 | 1.1.4 | Procurement of hardware | 10 days | 5 | 4/16 - 4/29 |
| 7 | 1.2 | Software Selection | 23 days | | 4/29 |
| 8 | 1.2.1 | Licensing negotiations with | 5 days | | 3/30 - 4/5 |
| 9 | 1.2.2 | Test software from each ve | 8 days | 8 | 4/6 - 4/15 |
| 10 | 1.2.3 | Procurement of software | 10 days | 9 | 4/16 - 4/29 |
| 11 | 1.3 | Integration of hardware and | 22 days | | 5/31 |
| 12 | 1.3.1 | Implement new security m | 5 days | 10,6 | 4/30 - 5/6 |
| 13 | 1.3.2 | Customization for each org | 7 days | 12 | 5/7 - 5/17 |
| 14 | 1.3.3 | Testing of the integrated d | 10 days | 13 | 3/18 - 5/31 |
| 15 | 1.4 | Verifying existing infrastruct | 20 days | | 6/28 |
| 16 | 1.4.1 | Test server capacity | 5 days | 14 | 6/1 - 6/7 |
| 17 | 1.4.2 | Test the network hardware | 5 days | 16 | 6/8 - 6/14 |
| 18 | 1.4.3 | Procure necessary infrastr | 10 days | 17 | 6/15 - 6/28 |
| 19 | 1.5 | Training for support team | 20 days | 14 | 6/1 - 6/28 |
| 20 | 1.6 | Create documentation mate | 30 days | | 7/12 |
| 21 | 1.6.1 | Technical architecture doc | 8 days | 14 | 6/1 - 6/10 |
| 22 | 1.6.2 | User manuals | 20 days | 21 | 6/11 - 7/8 |
| 23 | 1.6.3 | Review documentation ma | 2 days | 22 | 7/9 - 7/12 |
| 24 | 1.7 | Conduct user training | 12 days | 21 | 6/11 - 6/28 |
| 25 | 1.8 | Deployment by organization | 56 days | | 9/14 |
| 26 | 1.8.1 | Marketing deployment | 8 days | 24 | 6/29 - 7/8 |
| 27 | 1.8.2 | Engineering deployment | 8 days | 26 | 7/9 - 7/20 |
| 28 | 1.8.3 | Finance deployment | 8 days | 27 | 7/21 - 7/30 |
| 29 | 1.8.4 | Executives deployment | 8 days | 28 | 8/2 - 8/11 |
| 30 | 1.8.5 | Legal deployment | 8 days | 29 | 8/12 - 8/23 |
| 31 | 1.8.6 | IT deployment | 8 days | 30 | 8/24 - 9/2 |
| 32 | 1.8.7 | Administration deployment | 8 days | 31 | 9/3 - 9/14 |

**FIGURE 7.3**   First cut plan (partial).

limiting the project schedule. This means that any delay in these will cause the project to expand outward. Only tasks 17 and 18 do not show this code. Those tasks have *slack*, which means that they can be delayed some amount of time before becoming critical. The project summary bar shows a planned completion date of 10/14, and the other phase summary bars provide similar calculated completion values.

The advantages of using a software utility have already been pointed out, but here is a brief summary:

1. Calendar calculations respect non-working times.
2. Accurate translation of duration estimates to elapsed times.
3. Simplifying plan change process.
4. Calculating the critical path and slack times.
5. Producing status parameters beyond the basic view.

Figure 7.3 cuts off IDs higher than 32 to save space, but the key plan concepts are visible here.

Here is an interesting historical sidenote. When MS Project was first released, the goal was to show the plan as a graphical network of tasks, but the user population would only take the result in Gantt format, so the tool had to be redesigned to keep the old look intact. This is a classic example of how hard it is to change project culture.

The term *first cut* nomenclature used here means that more iterations will be needed to resolve differences among the various defined parameters and stakeholders—resources, sponsor, users, technical, financial, etc. For example, during the plan review discussions, it is decided that ID 6 should be 15 days rather than 10. In many such discussions, the team is trying to evaluate the impact of various changes. The software is capable of identifying this and corresponding resource gaps that affect the plan. At this point, the resulting plan should be presented to management and key stakeholders for final approval.

Several planning insights are embedded in this example. Note that the approved scope is still defined by the WBS and WP. Mechanically, this process offers the following results:

1. Tasks are scheduled according to the work calendar, and the data shown obeys non-work days. In this manner, the software integrates the task duration with the planned calendar work dates.
2. Bold black bars represent non-WP summary groupings of various types. Each of these has calculated finish dates based on lower-level tasks and attached to the bar.
3. The project critical path is denoted by the black hashed bars (red in color format). All of the critical path tasks constrain the schedule duration. Slippage of any critical path task will cause the project plan to slip by that amount.
4. Slack tasks do not constrain the schedule. These are illustrated by IDs 17–19.
5. Other status data is calculated by the software and available for presentation but not shown in this example.

# KEY MANAGEMENT VARIABLES

Two important management variables have emerged in this example: *critical path* and *slack*. Understanding both terms is vital for effective time management and requires that the project manager monitor these parameter values as the project unfolds. The critical path becomes a prime focus for schedule management. In theory, if you manage the critical path effectively, the project will finish according to the defined schedule, but any overrun on these tasks will have a negative impact on the schedule. Slack tasks simply mean that they are not constraining the output schedule, so usually will get less control interest.

The term *first cut* introduced in this chapter is a subtle but important idea in the management process. Think of the planning process as involving many players with different interests (i.e., schedule, resources, cost, functionality, etc.). Each of these players needs to be involved in the schedule approval process and this may require multiple iterations (cuts). As these decision iterations occur, the original scope definition may change some and the first cut plan will also need to follow those changes. At some point along this evolving trail, it will be time to stop the changes and say, "this is the approved plan." Our vocabulary term for that event is to *baseline* the plan. This term can be attached to any tracking variable such as schedule, budget, product function, or any other management variable. It says that this value is the approved goal for that variable. We will show more usage of this idea as we move through the life cycle.

## WBS UTILITY

The important role of the WBS in the overall management process cannot be overstated. This coded structure makes it possible to ensure that scope definition, WPs, and resulting plans are kept in step with each other. So, generating a WBS out of a software-created plan helps confirm that linkage integrity has been maintained properly. In other words, the current view did not have data entry errors. If a utility can generate a graphical WBS directly from the planning software, this linkage can be evaluated easily. A worthwhile utility to do this is WBS Schedule Pro, which is an add-in utility that generates WBS schematics directly from a Microsoft Project plan. Predecessor coding errors are easy to make, and they can be visually seen from this type of output. Figure 7.4 shows a sample WBS Schedule Pro output to highlight how various utilities of this type serve as useful analysis and communication support tools.

It is easy for data entry errors to slip by quietly if not analyzed in the format shown here. As with all computer-generated solutions, *"garbage in, garbage out."* This example also illustrates that the era of manual project management has been left behind. Communication needs and task dynamics simply make manual methods ineffective. Another valuable aspect of this approach is to produce more professional output. *There is value for the project team in image-building as well.*

We restate that a project plan view can be produced manually, but our bias is that any task that can be off-loaded to a computer is worth doing. The next chapter will follow this process to describe the financial planning process.

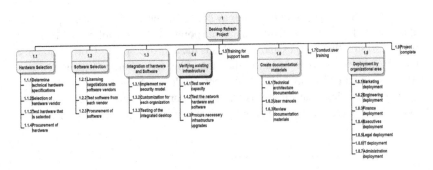

**FIGURE 7.4**    Expanded example WBS.

# MICROSOFT PROJECT REFERENCE SOURCES

For readers who would like to explore Microsoft Project software mechanics related to building the plan, a sample list of sources is listed below, and additional sources can be found on the Internet:

1. A full description of the plan construction mechanics process can be found in Carstens and Richardson (2020).
2. Cody Baldwin offers a free YouTube 15-minute overview of the mechanics that track pretty close to the text example. Available at https://www.youtube.com/watch?v=rWxUX2So-H4.
3. Melinda Schultz offers a free YouTube 55-minute video course that covers a broader view of the product. The video title is (2178) Scheduling for Success with Microsoft Project – YouTube.
4. A more extensive video course is available for a small fee at https://www.udemy.com/course/microsoft-project-tutorial/.
5. Several sources offer a summary MS Project keystroke cheat sheet. Search the Internet for MS Project Cheat Sheet.
6. A free limited-size trial version of WBS Schedule Pro and an online tutorial can be found at www.criticaltools.com. This utility generates WBS charts both internally and embedded in MS Project.

# SUMMARY

This chapter followed the WBS and WP scope definition view into a project schedule plan. Once estimates are created for each WP, that data is used in the scheduling process, culminating with an approved and baselined schedule. The value of using commercial software to handle the various construction mechanics is highlighted here. Also, the value of the schedule calculation software is highlighted. As projects grow beyond 30 or more tasks, managing the planning and control processes manually becomes untenable. Software and reusable templates are recognized as productivity levers that need to be utilized to speed up this process.

# REFERENCES

Carstens, D. S., and Richardson, G. 2020. *Project Management Tools and Techniques,* 2*nd ed.* Boca Raton, FL: CRC Press.

Critical Tools. 2020. WBS schedule pro. Available at: www.CriticalTools.com (accessed May 5, 2021).

# Project Budgeting

# 8

## INTRODUCTION

This chapter focuses on the processes related to developing a project budget, which builds directly from the Scope and Time chapters processes. The key focus remains on Work Breakdown Structure (WBS) and Work Package (WP) descriptions. We will now see how it is possible to map resource needs to work targets and look for resource availability gaps. From a pure arithmetic viewpoint, this is a simple calculation. If the required resources are defined for each WP, the estimated cost of each WP can be calculated. Beyond these simple raw calculations, there are other organizational aspects to the budgeting process. Also, it is now recognized that these estimates have a high potential to vary from the plan.

The third major step in the planning process involves creating a direct WP budget linked to the previous scope and schedule decisions. The Holy Trilogy (or Iron Triangle if you prefer that term) of project management is scope, schedule, and budget; these are the premier management topics most associated with a project. It is important to recognize that the budget will have the most visible external evaluation of the three because professional "dollar counters" have extensive systems to capture costs, which will be compared to project plans over time. Project managers who can manage these three variables successfully while producing the defined output product are very valuable to the organization.

The first step in our discussion here is to produce a first cut budget integrated with the scope and schedule described earlier. At this point, the focus moves to the associated resource data estimated for the WPs. This basic process involves "dollarizing" the planned resources for each WP using work estimating data. There are multiple resource categories involved in this process, but direct resources, material, indirect resources, and overhead are common groups.

Both schedule and budget estimates are negatively affected by various situations in the project environment. Sample sources outlined below can create variances for planned outcomes:

DOI: 10.1201/9781003218982-8

1. Project scope can grow from the baseline set, yet the approved budget and schedule remain constant. This technically should not be considered an overrun.
2. Estimates were inaccurate, and subsequent actuals are different. Are the variances the result of poor management, poor productivity, or just a bad estimate?
3. Resource quantity or skills availability did not match the plan, which then affects the schedule and budget. Tracking planned versus actual resource availability is a mandatory management view. If all other plan elements are accurate, this variance can explain why the project is not meeting the plan.
4. Project risk assessment was either poorly done or not done at all. Failure to consider events in this category can have a major impact on project outcomes. This management area is often ignored and not properly dealt with.
5. Unplanned events will occur, and they can have adverse effects on performance variables. There is certainly no magic technique to identify all of these, but it is wise to have some project reserve set aside to handle them. One example of this is weather events in a construction project environment. All plan factors can be in place in this situation, but the project does not meet the planned schedule because of Mother Nature.

One way to look at this type of list is to understand *Murphy's Law—Something unplanned will go wrong and at the worst possible time.* This summarizes the project world, and the project manager must understand it to survive. Before we try to deal with this dynamic environment, let's first review some basic mechanics related to the planned portion of the project and use that to create a first cut budget. Remember, after this step, a key stakeholder review iteration process takes place to gain formal approval. This review may be held concurrent with the plan schedule review described in the previous chapter.

## Direct Work Package Budget Process

It is important to restate that schedules and budgets must be created as part of a formal scope definition process at an appropriate level of detail. This statement's key theme is *that you cannot estimate or control the resulting process if you don't define the required work.* If you do find yourself the manager of Project Titanic, do not expect anything less than a negative outcome. Any other process breeds chaos and unsuccessful outcomes. Remember, the project team is the best source for defining the project parameters and not senior management or most other outside sources. *If the project manager does not protect the project plan integrity from various organizational group actions, it is best to get out before the volcano erupts.*

If the technical definition of a WP has been estimated reasonably well with the required resources (i.e., hours, skill levels, and resource rates), the computation of direct cost is simple arithmetic. The variability issues outlined above will need to be dealt with later, but the straightforward arithmetic version will be our goal for this first pass.

*Example calculation: WP X is estimated to require 80 hours of work by a resource making $30/hour. The calculated direct cost of that unit would be $2,400 (80 × 30). If material is also required for this work, that estimate would be added to the total, and this is repeated and summarized for any other defined resources.*

This calculated sum will be the direct cost for the WP, and we are defining that as the first cut value. For this budget to have organizational accuracy, it must also recognize various overhead costs for HR, material, and other sources. Every organization has a unique calculation factor, but be aware that this somewhat hidden cost is easily 100% of the direct cost. Many project managers resist this extra allocation of costs, saying that it is too much or not fair (like taxes). Here is a memory device to help understand why there are other costs in this calculation:

*A project is like a flower seed in the organizational flower bed. To make the flow grow, the organization supplies a gardener, dirt, fertilizer, and water. If all goes well, a beautiful flower emerges. Lack of any of these elements could mean less than desired flowers.*

The flower represents the project goal, and the flower bed represents the organizational overhead portion of the project.

At this point, the project will undergo scrutiny from various technical and organizational groups. Scope, schedule, budget, resources, and risk factors are reviewed. Financial constraints may be laid on the project plan requiring changes. After this review process is completed, the schedule and budget will be baselined, as described earlier in Chapter 7.

## Other Issues

At this point, other management issues begin to loom. For example, a planned resource isn't available in either quantity or skill level. This is often the number one reason for a task schedule or budget overrun. Failure to manage resource availability creates subtle but visible overruns even with good estimates. A resource gap can be created by either the needed resource not being available at all or not being available when needed. Managing both situations is essential to the management process. The vocabulary term for this activity is capacity management. Plan versus actual staffing variation is a key tracking variable. When this gap occurs, it should be anticipated that output variation will also occur. Timely corrective action is the management mantra here.

If the theoretical model for each WP was strictly followed, there would be detailed estimates for all input resources. As described in the previous chapter, Microsoft Project (or other similar project planning packages) will perform this arithmetic and assign the result to the plan WBS. Also, if this set of mechanics is followed, the software will now assess the status of resources for the project based on the assigned values. This process is called *resource capacity analysis*. Carstens and Richardson (2020) describe this software process in detail for the interested reader. This level of data capture is beyond what most organizations do; therefore, capacity analysis insight is lost, and resource gap surprise awaits later. The recommendation here is to perform detailed WP estimating so that the resource picture can be dealt with. This will give an advanced view of potential resource gaps and allow time to respond to those.

The more common budgeting approach is to perform the cost calculations externally and simply enter the resulting total WP cost value without accompanying resource information. So long as a cost value is produced for each work unit in the WBS, a project cash flow can be calculated. This shortcut produces a gap in the resource integrity chain and provides some viable insight into the project budget cash flow.

The discussion of resource capacity analysis practice shortfall brings up another term related to project management. We label this *Maturity*. If the goal is to produce successful projects and the topics described here are related, one must understand what a process gap adds to the project's risk level. In this example, not matching resources to tasks in the plan now leave the team in the dark regarding the upcoming resource picture. If project software is not used for this, some other technique needs to be defined. The recommendation is to use the integrated software as designed to help with the resource visibility issue, but one possible alternative is to create an operational resource tracking system that will manually assess the short-term need for resources, say for a month in advance. That still does not offer higher-level resource planning, but it is better than ignoring the problem completely.

# FIRST CUT CALCULATED COST

For this section, we will ignore resource gap issues (i.e., all required resources are available). To simplify this example, we will assume that each WP is estimated to

| ID | WBS | Name | Duration | Cost | | | | | | | | | | | | |
|----|-----|------|----------|------|---|---|---|---|---|---|---|---|---|---|---|---|
| 1 | 1 | ◢ Desktop Refresh Project | 151 days | $25,000 | | | | | | | | | | | | ● $25,000 |
| 2 | 1.1 | ◢ Hardware Selection | 23 days | $4,000 | | ● $4,000 | | | | | | | | | | |
| 3 | 1.1.1 | Determine technical hardw | 4 days | $1,000 | | ▣ $1,000 | | | | | | | | | | |
| 4 | 1.1.2 | Selection of hardware ven | 1 day | $1,000 | | ▌ $1,000 | | | | | | | | | | |
| 5 | 1.1.3 | Test hardware that is selec | 8 days | $1,000 | | ▬ $1,000 | | | | | | | | | | |
| 6 | 1.1.4 | Procurement of hardware | 10 days | $1,000 | | ▬ $1,000 | | | | | | | | | | |
| 7 | 1.2 | ◢ Software Selection | 23 days | $2,000 | | ● $2,000 | | | | | | | | | | |
| 8 | 1.2.1 | Licensing negotiations with | 5 days | $1,000 | | ▣ $1,000 | | | | | | | | | | |
| 9 | 1.2.2 | Test software from each ve | 8 days | $1,000 | | ▬ $1,000 | | | | | | | | | | |
| 10 | 1.2.3 | Procurement of software | 10 days | $0 | | ▬ $0 | | | | | | | | | | |
| 11 | 1.3 | ◢ Integration of hardware and | 22 days | $3,000 | | | ● $3,000 | | | | | | | | | |
| 12 | 1.3.1 | Implement new security m | 5 days | $1,000 | | | ▬ $1,000 | | | | | | | | | |
| 13 | 1.3.2 | Customization for each org | 7 days | $1,000 | | | ▬ $1,000 | | | | | | | | | |
| 14 | 1.3.3 | Testing of the integrated d | 10 days | $1,000 | | | ▬ $1,000 | | | | | | | | | |
| 15 | 1.4 | ◢ Verifying existing infrastructu | 20 days | $3,000 | | | | ● $3,000 | | | | | | | | |
| 16 | 1.4.1 | Test server capacity | 5 days | $1,000 | | | | ▬ $1,000 | | | | | | | | |
| 17 | 1.4.2 | Test the network hardware | 5 days | $1,000 | | | | ▬ $1,000 | | | | | | | | |
| 18 | 1.4.3 | Procure necessary infrastru | 10 days | $1,000 | | | | ▬ $1,000 | | | | | | | | |
| 19 | 1.5 | Training for support team | 20 days | $1,000 | | | | ▬ $1,000 | | | | | | | | |
| 20 | 1.6 | ◢ Create documentation mater | 30 days | $3,000 | | | | | ● $3,000 | | | | | | | |
| 21 | 1.6.1 | Technical architecture doc | 8 days | $1,000 | | | | | ▬ $1,000 | | | | | | | |
| 22 | 1.6.2 | User manuals | 20 days | $1,000 | | | | | ▬ $1,000 | | | | | | | |
| 23 | 1.6.3 | Review documentation ma | 2 days | $1,000 | | | | | ▮ $1,000 | | | | | | | |
| 24 | 1.7 | Conduct user training | 12 days | $1,000 | | | | | ▬ $1,000 | | | | | | | |
| 25 | 1.8 | ◢ Deployment by organization | 64 days | $7,000 | | | | | | | ● $7,000 | | | | | |
| 26 | 1.8.1 | Marketing deployment | 8 days | $1,000 | | | | | ▬ $1,000 | | | | | | | |
| 27 | 1.8.2 | Engineering deployment | 8 days | $1,000 | | | | | ▬ $1,000 | | | | | | | |
| 28 | 1.8.3 | Finance deployment | 16 days | $1,000 | | | | | ▬ $1,000 | | | | | | | |
| 29 | 1.8.4 | Executives deployment | 8 days | $1,000 | | | | | | ▬ $1,000 | | | | | | |
| 30 | 1.8.5 | Legal deployment | 8 days | $1,000 | | | | | | ▬ $1,000 | | | | | | |
| 31 | 1.8.6 | IT deployment | 8 days | $1,000 | | | | | | ▬ $1,000 | | | | | | |
| 32 | 1.8.7 | Administration deployment | 8 days | $1,000 | | | | | | ▮ $1,000 | | | | | | |

**FIGURE 8.1**   First cut budget view.

cost $1,000 direct cost. This value is shown as data in the column titled Cost and calculated by the software from estimated resource data for the WP. As before, note that the software knows the WBS structure grouping and correctly calculates phases and the total project (sum of the WP costs). Figure 8.1 uses the costing parameters to calculate the total project direct cost to be $25,000.

This calculation may well be good enough for many projects, but it does ignore the questions related to resource status, potential reserves needed, project overhead, and possibly other costing issues that should be considered.

Figure 8.2 shows the same WBS hierarchy as Figure 8.1, except cost data is now added to the view. This makes an excellent way to illustrate how costs are arrayed across the project.

# PROJECT BASELINE

Microsoft Project has the functionality to define multiple baselines. For this discussion, we will assume that there is only one, the initially approved baseline. But the software can also display other approved plans. Figure 8.3 shows how the baseline is presented on top of the active plan.

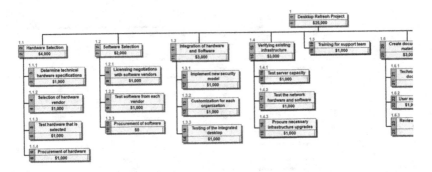

**FIGURE 8.2**   WBS with cost labels.

Note that two bars are now shown for each line item. The hashed bar represents the original frozen baseline (or possibly an alternative baseline), and the solid bar is the current plan. The actual duration bars will move as the project evolves. This two-state view provides a visual comparison to show how the project compares to a particular baseline. This type of presentation is often used to show project status. Similar textual data can also be added to the spreadsheet data view or on the bars to aid in status analysis. In practice, showing baseline comparisons to actual performance is an uncomfortable idea for a project team, but one that should be followed in some formal manner. The philosophy here is to view this type of status as "an honest view of the truth." Status should always be linked to an approved baseline. There can be a lot of emotion over this topic as the project team often feels like they were not the cause of the gap, yet they are blamed for the overrun. It is these gaps that will consume a lot of management time in evaluating and explaining.

## Scope Creep

One related team management area of concern is recognizing that factors such as changing scope can make fixed baseline versus actual comparisons show inaccurate status. Status should be measured by a historic metric value that recognizes this class of change. Here is a simple example. It is not uncommon for scope changes to be 25% above the approved plan. Management approves this new scope, yet the original baseline was created before this and did not change. If the plan now shows an overrun of 25%, should this increase be viewed as a project overrun reflecting poorly on the project team? Obviously not! The correct procedure should be to show an expanded scope and use that as a performance measure. This is now a different project. Nevertheless, it is interesting to evaluate how much the project scope does *creep* over time

| ID | WBS | Name | Duration | Cost | | Half 2, 2021 |
|---|---|---|---|---|---|---|
| 1 | 1 | Desktop Refresh Project | 151 days | $25,000 | | $25,000 |
| 2 | 1.1 | Hardware Selection | 23 days | $4,000 | | $4,000 |
| 3 | 1.1.1 | Determine technical hardware specifications | 4 days | $1,000 | | 4/2 |
| 4 | 1.1.2 | Selection of hardware vendor | 1 day | $1,000 | | 4/5 |
| 5 | 1.1.3 | Test hardware that is selected | 8 days | $1,000 | | 4/15 |
| 6 | 1.1.4 | Procurement of hardware | 10 days | $1,000 | | 4/29 |
| 7 | 1.2 | Software Selection | 23 days | $2,000 | | $2,000 |
| 8 | 1.2.1 | Licensing negotiations with software vendors | 5 days | $1,000 | | 4/5 |
| 9 | 1.2.2 | Test software from each vendor | 8 days | $1,000 | | 4/15 |
| 10 | 1.2.3 | Procurement of software | 10 days | $0 | | 4/29 |
| 11 | 1.3 | Integration of hardware and Software | 22 days | $3,000 | | $3,000 |
| 12 | 1.3.1 | Implement new security model | 5 days | $1,000 | | 5/6 |
| 13 | 1.3.2 | Customization for each organization | 7 days | $1,000 | | 5/17 |
| 14 | 1.3.3 | Testing of the integrated desktop | 10 days | $1,000 | | 5/31 |
| 15 | 1.4 | Verifying existing infrastructure | 20 days | $3,000 | | $3,000 |
| 16 | 1.4.1 | Test server capacity | 5 days | $1,000 | | 6/7 |

**FIGURE 8.3**   Baseline plan with finish dates (partial WBS).

because of approved scope changes. Conceptually, these changes represent new baselines, but that is infeasible to track. Regardless of the comparison methodology used, beware not to interpret performance using factors outside of the team control, such as the scope change situation outlined here.

# MANAGING PLAN VARIABILITY

A common method of handling variability is simply adding padding to each task to cover the type of variations outlined here. This practice needs to be curtailed as it only acerbates the overrun issue. First, padding hides visibility into specific variations, but there is something even more damaging from this practice. Some organizations simply add 100% to an estimate to cover the potential variation. On the one hand, this artificially expands the plan duration and cost, but the behavioral aspect is even worse. There is a psychological issue related to this practice. If the worker has twice as long to perform a task as needed, guess how long you will take? At least twice as long! Actual practice shows that not only will it take twice as long, but it will even often expand more beyond that point. This phenomenon is more psychological than technical. There simply is no internal motivation to start work on a padded task using this practice since you are aware the task has extra time. Humans react to this accordingly—i.e., procrastinate until the last minute, just like we all used to do in school. This behavior even has a formal name called the *student syndrome*. Because of this behavioral phenomenon, the task estimating practice should avoid padding individual tasks. The recommended practice should be to estimate on a 50/50 basis—estimate the time and cost for a 50% probability it will overrun (or underrun). This means that the overall project will now have to have a reserve to handle statistical overruns. The important aspect of this approach is to give the team challenging time estimates with no extra time to procrastinate. All tasks now require an increased focus to finish per the reduced estimated values. This task estimating technique has been shown to cut 25% off project schedules and similar savings on budget when managed properly. Recognize that this type of scheduling strategy is not common but should be. This practice is cause by cultural bias to avoid overruns, so it is important to change that approach to work—not an easy one to fix for sure.

*Authors note: The concept of 50/50 scheduling is likely the most complex management item described in the book and will be challenging to implement, but the potential impact on project results is too significant for it to not be included here. Just realize what traditional padding does to the task measurement accuracy and associated status accuracy. The traditional approach to estimating leads to overruns as a result of the procrastination tendency, and in the end, you have no clue as to what time that task should have taken. The cycle then repeats using false historical data.*

A companion second part of the process involves how to cover from the 50/50 strategy. The answer to that is to use a project-level schedule and cost buffer. Introduction of these buffers introduces a new control variable into the model. Recognize that every management strategy solution contains some counter-issue to resolve. In this case, the emergence of a formal reserve fund in the plan is not linked to a defined WBS box and, for this reason, is often viewed as padding (which it is, but more organized and visible). It is now necessary for the project manager to explain this logic to management and project team members. This must be explained as a logical contingency approach to overruns. Recognize that there is a risk here that senior management may say to take this out of the plan since it is phony, but omitting this triggers a real overrun problem. The project manager challenge becomes one of educating the various external stakeholder groups regarding how the reserve will be handled in practice. Reserves of this type are added to the project schedule and budget by attaching a dummy task (buffer) containing the amount of schedule and cost reserve to be allocated. An example of a schedule reserve is shown in

Figure 8.4, which shows a buffer added to the end of the bike project plan introduced in Chapter 7.

Note the "dummy task" added to the structure as ID 30 with a duration of 20 days. That added buffer extends the plan project completion by that amount to now show 3/18. The tricky question remaining is how 20 days got selected. One way to do this is to have collected data from previous projects and use that as an estimation guide. Also, a good risk assessment process could be used to help assess potential variability. In addition, there are techniques available to model projects using simulation to develop a probabilistic range of completion dates. More details on this are available in Chapter 24 in Carstens and Richardson (2020).

Regardless of the method chosen, buffers will be questioned as to their legitimacy and size. Realize that they are needed to deal with reality, and they will need to be discussed in various operational meetings. For example, if the project is 50% of the way through the life cycle and has used 70% of the buffer, there is the need to evaluate what this means to the projected completion date. Unless this overrun trend can be turned around, the project is at risk of a schedule (or budget) overrun. Buffers can also be added to plan phases in place of a single project buffer.

*Note: When you show a project plan with a named completion date, you can be almost 100% sure that this date is not accurate. Get your stakeholders used to range values instead. This is like a long-term weather forecast saying it will rain at noon on some future date.*

Project plans are roadmaps based on complex estimates and assumptions. It is much better to get management and stakeholders to understand the reasons for variability. This is just one of the naiveties of project management that needs to be better understood by all. Proper use of reserves is a valuable technique to introduce this idea.

# RESOURCE MANAGEMENT MECHANICS

This section will describe the high-level mechanics for linking resources to the cost calculation. As stated earlier, the value of doing this inside the planning software helps to achieve integration between the plan elements. In this fashion, the cost will be calculated using the internal resource data, and the results can be matched to defined resource levels. Even though this mechanical

| WBS | Task Name | Duration | Pred. | Oct | Nov | Dec | Jan | Feb | Mar |
|---|---|---|---|---|---|---|---|---|---|
| 0 | Bicycle | 118 days | | | | | | | 3/18 |
| 1 | Vision Documenation | 13 days | | 10/22 | | | | | |
| 1.2 | System Design | 10 days | | 10/17 | | | | | |
| 1.3 | Define Bill of Materials | 3 days | 3 | 10/18 — 10/22 | | | | | |
| 2 | Frame Set | 25 days | | | 11/26 | | | | |
| 2.1 | Frame | 5 days | 4 | 10/23 — 10/29 | | | | | |
| 2.2 | Handlebars | 5 days | 6 | 10/30 — 11/5 | | | | | |
| 2.3 | Fork | 5 days | 7 | | 11/6 — 11/12 | | | | |
| 2.4 | Seat | 5 days | 8 | | 11/13 — 11/19 | | | | |
| 2.5 | Unit Test | 5 days | 9 | | 11/20 — 11/26 | | | | |
| 3 | Crank | 10 days | 10 | | 11/27 — 12/10 | | | | |
| 4 | Wheels | 10 days | | | | 12/24 | | | |
| 4.1 | Front wheel | 5 days | 11 | | | 12/11 — 12/17 | | | |
| 4.2 | Rear Wheel | 5 days | 13 | | | 12/18 — 12/24 | | | |
| 5 | Braking System | 10 days | 14 | | | 12/25 — 1/7 | | | |
| 6 | Suppor Takss | 50 days | | | | | | | 3/18 |
| 6.1 | Shifting System | 10 days | 15 | | | | 1/8 — 1/21 | | |
| 6.2 | System Assembly | 10 days | 17 | | | | 1/22 — 2/4 | | |
| 6.3 | System Test | 10 days | 18 | | | | | 2/5 — 2/18 | |
| 7 | Project Buffer | 20 days | 19 | | | | | 2/19 Buffer | 3/18 |

**FIGURE 8.4**   Project plan with completion buffer.

process is cumbersome, it potentially provides a good resource evaluation process that provides time to correct defined gaps.

## Resource Sheet

The first step of the costing process is to define the available resource profile. This is the assumed resource pool available to the project. Figure 8.5 shows the Microsoft Project Resource Sheet data elements for defining the available resource pool.

A brief definition for each field follows:

Resource name—in this example, the resources are defined by function rather than by specifically named workers (generic names and rates can be used here)

Type—the type of resource (i.e., worker, material, or dollar)

Material—used for pricing material

Initials—shorthand reference to the resource

Group—worker's organizational home

Max—maximum number of that resource (200% means two individuals)

Std. rate—costing parameter (i.e., hours or days rate)

Ovt. rate—how to incrementally cost if the worker is assigned more than regular hours

Cost/use—a setup-type variable (i.e., installing a scaffold, renting a crane)

Accrue at—when to assign the cost (start, end, or prorate over the assigned time)

Calendar—which calendar is to be used for availability (project, individual)

## Base-level Plan

Figure 8.6 shows the sample plan that will be used to demonstrate the costing process.

The following are three new data columns opened in this view:

Resource names—this shows the assigned resources to each task

Cost—this is the computed cost based on the allocated resources

Information (column 1)—resource status flag (more on this below)

Also, there is one new highlight flag that emerges from this process. Note the "Red Man" icons shown in column 1. When this information flag occurs, it signifies that the task shown cannot be supported by the resources defined for that task. This is an early warning to the project manager that corrective action is needed. Here we see another example of added value supplied by using the calculation model as illustrated. Knowing this gap situation, several replanning techniques can be used to bring the plan back into balance. The simplest solution to this situation is adding more resources to the pool, but this may not be feasible. In that case, several other options will need to be reviewed. Carstens and Richardson (2020) offer a good overview of these options and the related mechanics needed to resolve the resource gap issue. In operational mode, this is one of the most complex analytical processes in the project management model. To date, an automated solution to this class of problem has

| Resource Name | Type | Material Label | Initials | Group | Max. Units | Std. Rate | Ovt. Rate | Cost/Use | Accrue At | Base Calendar |
|---|---|---|---|---|---|---|---|---|---|---|
| Designer | Work | | DES | DES | 400% | $75/hr | $75/hr | $0 | Prorated | Standard |
| Tester | Work | | TEST | TEST | 200% | $45/hr | $45/hr | $0 | Prorated | Standard |
| Business Analyst | Work | | BA | BA | 500% | $50/hr | $50/hr | $0 | Prorated | Standard |
| Mechanical Engineer | Work | | ME | ME | 400% | $40/hr | $40/hr | $0 | Prorated | Standard |
| Electrical Engineer | Work | | EE | EE | 200% | $35/hr | $35/hr | $0 | Prorated | Standard |
| Procurement | Work | | PROC | PROC | 500% | $25/hr | $25/hr | $0 | Prorated | Standard |
| Mechanic | Work | | MECH | MECH | 200% | $40/hr | $40/hr | $0 | Prorated | Standard |
| Driver | Work | | DRIVE | INV | 100% | $100/hr | $0/hr | $0 | Prorated | Standard |
| Fiberglass | Material | | GLASS | INV | | $225 | | $0 | Prorated | |
| Tires | Material | | TIRE | INV | | $50 | | $0 | Prorated | |
| Wires | Material | | WIRE | INV | | $300 | | $0 | Prorated | |
| Metal | Material | | METALS | INV | | $0 | | $0 | Prorated | |

**FIGURE 8.5**   Resource sheet.

evaded software solutions; however, the current software capability does offer some calculation help in evaluating various options.

## Cost Calculation

The process of computing a task cost starts with the WP information. In Microsoft Project, by right-clicking on the task in the plan view, a resource allocation box pops up. Figure 8.7 shows this box allocated with costing data from a WP. A business analyst, designer, electrical engineer, and mechanical engineer are allocated at 100% each, resulting in the WP direct cost being automatically calculated. *Note:* This data has to be entered manually for each task.

The calculation process uses stored resource and plan data. First, the plan knows that the task is five days in duration, and the resource costs are matched to the names allocated. Finally, the work calendar time periods for the cost to occur. The WP cost is summarized as $45.600 and properly slotted into work days from this collection of data. This same set of mechanics is used for all the tasks, and a computed direct budget is aggregated into line 1 of the project plan. Once again, this points out how the process integrates with the various planning parameters from scope definition and schedule to now producing a cash flow calculation for the project plan.

| | WBS | Task Name | Predecessc | Duration | Cost | Resource Names |
|---|---|---|---|---|---|---|
| 1 | 1 | ⊿ Car Problem First Cut | | 501 days | $1,252,920 | |
| 2 | 1.1 | ⊿ Planning & Project Initiation | | 75 days | $166,400 | |
| 3 | 1.1.1 | Charter | | 10 days | $17,200 | Business Analyst ,Designer |
| 4 | 1.1.2 | Stakeholders Identification | 3 | 5 days | $18,200 | Business Analyst |
| 5 | 1.1.3 | Scope Definition | 3 | 15 days | $45,600 | Business Analyst ,Designer,Electrical Engineer,Mechanical En |
| 6 | 1.1.4 | Schedule Development | 5 | 15 days | $50,400 | Designer,Electrical Engineer,Mechanical Engineer |
| 7 | 1.1.5 | Risk Assessment | 6 | 5 days | $8,600 | Business Analyst ,Designer |
| 8 | 1.1.6 | Budget Definition | 7 | 15 days | $11,400 | Business Analyst |
| 9 | 1.1.7 | Management Charter Approval | 8 | 5 days | $3,800 | Business Analyst |
| 10 | 1.1.8 | Set baseline | 9 | 5 days | $11,200 | Business Analyst |
| 11 | 1.2 | ⊿ Engineering | | 286 days | $737,460 | |
| 12 | 1.2.1 | ⊿ Body / Engine Draft Design | | 45 days | $149,400 | |
| 13 | 1.2.1.1 | Initial Draft | 9 | 35 days | $121,800 | Designer[200%],Electrical Engineer[200%],Mechanical Engine |
| 14 | 1.2.1.2 | Integration analysis | 13 | 10 days | $27,600 | Electrical Engineer[200%],Mechanic [200%],Mechanical Engi |
| 15 | 1.2.2 | ⊿ Mechanical Engineering | | 110 days | $238,800 | |
| 16 | 1.2.2.1 | Engine Design | 10 | 80 days | $182,400 | Designer[200%],Mechanical Engineer |
| 17 | 1.2.2.2 | Mechanics Design | 10FS+10 days | 30 days | $56,400 | Designer,Mechanical Engineer [300%],Mechanic |
| 18 | 1.2.3 | ⊿ Electrical Engineering | | 72 days | $148,080 | |
| 19 | 1.2.3.1 | Electrical design | 17 | 30 days | $81,600 | Designer[200%],Electrical Engineer[300%] |
| 20 | 1.2.3.2 | Wiring | 19 | 15 days | $27,600 | Electrical Engineer[200%],Designer |
| 21 | 1.2.3.3 | Others Elect | 19 | 27 days | $38,880 | Electrical Engineer,Mechanic [200%] |
| 22 | 1.2.4 | ⊿ Structural Engineering | | 44 days | $120,480 | |
| 23 | 1.2.4.1 | Body | 12 | 30 days | $66,080 | Designer,Mechanic ,Mechanical Engineer |
| 24 | 1.2.4.2 | Other Body Parts | 12 | 30 days | $43,200 | Designer,Procurement |
| 25 | 1.2.4.3 | Design integration review | 14,17,21,24 | 10 days | $11,200 | Procurement |
| 26 | 1.2.5 | ⊿ Body / Engine Structure | | 30 days | $69,500 | |
| 27 | 1.2.5.1 | Chassis | 23 | 20 days | $24,450 | Metal [3.5],Mechanic [300%] |

**FIGURE 8.6**   Sample project plan.

## Resource Reports

Numerous text and graphical reports can be produced at this point, further supporting the value of using software for the planning and tacking processes. Through this activity, a wide variety of data can be extracted with minimal extra work.

## Comments on the Costing Process

Recognize that all model processes are abstractions of reality by definition and the costing model calculation is no exception. Here are a few sample observations regarding the possible inaccuracy of this process:

1. The actual resource that performs this work will likely have a different rate than the one shown. So, a cost variance will be created by the rate difference. That situation deserves analysis in the tracking process.
2. If a WP were to be larger than recommended (e.g., 80 hours), the timing computed for the resource would be affected. Also, each resource might not work continuously, so their actual charges would be phased differently.

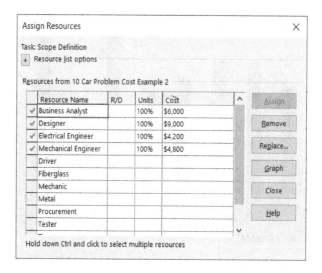

**FIGURE 8.7**   Task 1.1.3 resource allocation.

3. Capturing material charges is often not feasible at a WP level. It may be so infeasible that all material will be handled external to the software plan.

4. Schedule slippage changes the actual resource picture to the degree that the maximum resource level may be exceeded even if the model does not show that.

Given the value of cost and resource status related to this process, it is recommended to be followed in practice even if some shortcuts are required because of the particular project environment.

## PROJECT CASH FLOW

Financial forecasting is close relative to the resource analysis picture. Figure 8.8 shows a sample computer-generated report showing project cash flow. Reports of this type can show either the current plan values or some comparison of the current plan versus baseline values.

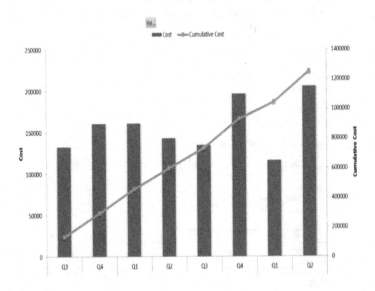

**FIGURE 8.8**   Cost report.

# RESOURCE MANAGEMENT

As stated earlier, resources play a pivotal calculation role in both schedule and budget data generation. In the ideal situation, the project team would have proper resources standing by when it was time to start a task. Resource availability gaps are a common management issue. Looking at this from a project management improvement viewpoint, resolving resource availability gaps would likely do more to make the project finish on schedule than any other execution stage management strategy. Resource gaps can produce an overrun even if the project is perfectly planned. The project manager is charged with successful completion, yet in many cases does not own the required resources. If this is the environment, resource availability status must be carefully tracked as the root cause for overruns. This may include both the quantity and quality of the resources. If this is not done, the external stakeholder conclusion will be that the team did not perform. One of the key roles of a project manager is to protect their team from unwarranted criticism, and this is certainly a situation to monitor.

This abbreviated text format does not allow space for describing detailed mechanics regarding how to formally allocate resources to WPs in a mechanized software plan, but the references cited earlier will help with that. Admittedly, this is a time-consuming exercise, but it should be considered if resources are limited. One alternative is to work out a scheduling and tracking system to help ensure that resources are available as needed. In some cases, the problem is more scheduling than a true availability gap. Human resources tend to be the key operational resource constraint, and this issue cannot be ignored. Whenever a resource is not available, and this is known in advance, the project schedule needs to be adjusted. If buffers are used in the plan, they can be appropriately used to keep the completion schedule intact.

Several budget-related planning issues have been mentioned in previous sections. Let's take a moment to review these as a group and comment on the budgeting aspect of each.

*Scope Variation.* Project scope can change in both formal and informal ways. Think of this as new WPs are added into the mix. Each of these additions brings additional work-related schedule and budget implications that can create overruns. All scope changes must be managed carefully.

*Unplanned Events.* This topic is more complex than discussed here. Was this something that could have been anticipated, or did it just suddenly emerge with no warning? More details regarding a formal risk analysis process will be presented in Chapter 9. For now, simply recognize that various events will occur during execution that is not defined in the approved plan. They will likely negatively impact the planned project result if not handled by the risk management process. This is another area where buffers are used to protect the result.

*Overhead.* This topic can be quite complex in large organizations, but it is a planning issue that must be dealt with in every project. This term is primarily related to the budget calculation and related management process.

*Level of Effort.* Nothing has been said about this type of task thus far, but this class of resource may consume significant project resources. For instance, project managers are often charged with the project budget, but they do not work specifically on a WP. This may consume 10% of the project budget. Other similar activities and resources fall into this same category (i.e., IT support, a copy machine, project car, administrative resources, etc.). All of these non-direct sources must be captured in the project budget and managed appropriately.

*Reserves.* These have been titled buffers thus far, but various other types of reserves are related to defined risks, scope change, and other variable categories.

# CONTROL ACCOUNTS

Up to this point, cost management has focused on items at the WP level; however, this sometimes is not the desired place to manage project activity. In this case, a closely commingled group of WPs is judged to be the better alternative. WPs can be grouped into some higher-level collection for management purposes. These groups are called Control Accounts (CAs) and are used for management and data collection points. The design concept for CAs is to collect actual status where it makes the most sense. One other consideration is the questionable accuracy of data collected at a lower WP level. The administrative work to capture data is yet another deciding factor. Many times, the selection logic involves numerical data accuracy versus operational cost. The one thing to consider in this decision is the value of the lower-level information versus how likely WP-level data is to be accurate. If one moves the collection point upward, visibility below that level is lost. Industry experience is now showing that having historical information is worthwhile for future planning, so the data collection level decision is more important than it appears to be at first glance.

Another related management trend in this area is to formally name a team member as delegated managers for specific control points. The emerging name for this is a CA Manager or a CAM. This is a form of delegating management authority and improving team commitment to the plan. Through this method, lower-level elements in the project team feel more responsible for the outcome. Formalizing CAs and defining CAM responsibility is an operational strategy worth considering.

## SUMMARY

The project budget defines the anticipated spending plan for the organization. This includes the total amount of dollars required and the categories and time frames for these expenditures. Accounting systems are very adept at capturing actual project resources consumed, and this data is more available than any other project data category. In addition, senior management is fully tuned to this category, and for that reason, it must be carefully planned and managed.

A cost-related term that is becoming somewhat recognized in the industry is the *Performance Measurement Baseline* (PMB). This formal budget level includes all the lower-level components plus some reserve allocation. This term serves as a modern baseline concept with very similar views described here. Also, recognize that baseline-oriented terms such as this can deal with more than schedule or budget. Technical organizations might define named baseline terms for their planned product performance goals (i.e., how fast, target weight, size, etc.).

Chapters 6, 7, and 8 have collectively highlighted the requirements for developing the project scope, schedule, and budget. These topics have historically been called the Holy Trilogy or Iron Triangle of project management. Industry practitioners have a core focus on these three variables, but not necessarily in an enlightened manner.

These chapters have described the key processes and general mechanics that should be followed to produce an integrated project plan that will support other processes through the life cycle. It is important to be reminded that a project is a dynamic activity, and the plan must react as necessary to provide the proper guidance and visibility. Future chapters will focus on other related aspects of this complex management environment. Up to this point, our goal has been to build a clear and simple foundation from which to move forward.

Despite these somewhat high-level views, this is a reasonable base point to expand from for future aspects of the overall management spectrum.

# REFERENCES

Carstens, D. S., and Richardson, G. 2020. *Project Management Tools and Techniques,* 2*nd ed.*. Boca Raton, FL: CRC Press.

Critical Tools. 2020. WBS schedule pro. Available at: www.CriticalTools.com (accessed May 5, 2021).

Project Management Institute (PMI). 2017. *A Guide to the Project Management Body of Knowledge,* 6th ed. Newtown Square, PA: PMI.

# Dealing with Project Risk

# 9

## INTRODUCTION

Dealing with project risk is a very complex issue in that neither the potential events that will occur nor the cost of those events is easily determined. Nevertheless, it is important to follow a model of assessment and handling of these events. Associated with this, the management side of the equation needs to develop a risk culture that helps to establish sensitivity to these variations in the project plan. The planning process will establish a contingency reserve designed to protect the project budget and schedule from risk events that emerge during the life cycle. Focus on this issue over time will improve the accuracy of the results.

It should be very obvious that every project has some form and degree of risk. These events are often hard to anticipate, and when they occur, the project plan will likely be negatively impacted regarding budget and schedule. For this basic reason, planning efforts should be undertaken to identify potential factors and minimize their impact on the project outcome. The process of managing and controlling this aspect of the project life cycle is still considered immature but is more widely recognized now than in the past.

This chapter will describe an industry-standard process of assessing project risk and show how this process is monitored in the execution phase. It is important to recognize that accurate mathematical quantification of the project risk is usually impossible, except in rare repeatable-type projects. However, industry experience indicates that organizations that focus on risk tend to have a less negative impact than those that ignore the topic. The general conclusion of this result is attributed to the organizational focus on risk more than quantification. Think of this as developing a *risk culture*. When the project team focuses on risk, they will deal with it better because *risk tracking* and general awareness become an operational focus. It is necessary to go through the risk assessment process and do a reasonable job of identifying those tasks that will potentially cause trouble and identify broader higher-level issues that may impact the project. The output of this will produce a database of identified risks and a method to monitor these targets during the life cycle. Because

DOI: 10.1201/9781003218982-9

risk events cause budget and schedule issues when they occur, a risk *reserve* should be established and a *risk map* created to identify the critical time frame for monitoring these events. Both of these management tools will be described in more detail later in the chapter.

The list below summarizes answers to common questions related to this topic:

*What is Risk?:* In the industry vernacular, one form of this is called a "known/unknown." In other words, it is something that you should suspect could happen but may not. The second form of risk is called an "unknown/unknown." Each risk type is considered to have different degrees of predictability. The unknown/unknown likelihood of occurrence is so low that it falls outside of the normal assessment range. An extreme example of this is a meteor falling on a building. Note that this is an uncontrollable risk and therefore only impacts the decision as to whether to pursue the project (accept or avoid). The known/unknown category can be more accurately predicted than this, and therefore it is the focus of the assessment process.

*How are risks identified?:* In most project environments, there is some background history that helps with identifying the things that can happen in the course of the life cycle. These are called *risk events*. In a new project type with no background history, the identification approach would require more creativity and expertise to identify risks accurately.

*What can be done about risk?:* The most important warning is to not ignore the topic and be blind to its occurrence. That is too often the case. If you realize that a risk event will hurt project performance, the obvious reaction is to try to mitigate its impact (i.e., remove it, decrease it, transfer it, and/or monitor it).

*How do I modify the plan as a result of this?:* More explanation is needed to answer this question. Philosophically, the approach to this phenomenon is to take a management-oriented process that covers both the planning process and later monitoring the ongoing outcome result through the life cycle. Think of this process as a management model. Let's reserve more on this answer and work on the process details in this chapter.

# RISK CATEGORIES

There are two types of risks. At the high end of the categorization hierarchy, the two classes are *Known* and *Unknown*. The known group falls into an assessment class that has the potential to occur and for which some general analysis of the probability and impact may be estimated. These risks will be dealt with by following the logical assessment model steps outlined below. The second category type, unknown/unknown risk group, falls into the domain of events that are not generally anticipated and generally fall outside of formal assessment. The value obtained from examining the project through a lens focused on how these probabilistic events can impact the project is a valuable step in improving the planned outcome.

Here are two examples of risk that help in understanding the thought process for this process area of the management problem. First, do you carry a spare tire in your car? Why do that? This is a simple example of a known/unknown. With this example, you are mitigating the impact of a flat tire occurring—not stopping it, but making it less onerous. We might add that the driver should know how to execute the triggered risk event (i.e., how to change the tire). A second example shows the characteristic of the unknown/unknown type and how the handling of this example results. The unknown risk example event occurs unexpectedly as a newly finished nuclear power plant starts up. At that point, two endangered spotted owls are found to be nesting in the exhaust stack. This was not included in the risk assessment process, so now the project has a defined panic-oriented risk-handling question—what to do with the owls? If you live in Texas, this is not going to be a long-term problem (the reader will have to interpret this statement). However, if a powerful environmental group is involved, the project may be shut down as a slow legal process unfolds, or until the owls decide to leave. If this event had been anticipated, a possible mitigation action would have helped the owls find an alternative home. In both of these examples, the situation was not as desired, and the impact was varied. That is the general view of planned versus unplanned triggered risk events— i.e., it happened, what to do now?

## RISK ASSESSMENT PROCESS

The first step in the risk management process is to document a *Risk Plan* used to guide the overall process. The major sections of this document define the following assessment areas:

Methodology to be used (i.e., tools, data sources, process, etc.)
Roles and responsibilities—assign resources and roles
Budgeting—establishes resource limits on the activity
Risk categories—establishes the overall structure and scope of the analysis

Quantification goals—establishes the desired level of quantification for the analysis

Risk documentation—specifies how the output will be documented and the role of the Risk Register

Using the risk plan as a guide, the goal of the risk assessment process is to identify and find corrective actions that will minimize the impact of these events. The formal risk assessment model is divided into subprocess groups designed to identify, prioritize, quantify, mitigate, and monitor the ongoing state. Realize that risk events can be envisioned before the project start, but new ones can also occur later, so this process must be looked at as a full life cycle activity and not just a planning process. An important segment of this process is to capture the various items in a data store titled the *Risk Register*, which will be used as a formal support document for the overall process.

# Identification Step

The primary goal of the identification process is to create a list of potential risk events that will be reviewed further in the next step. Historical experience with a similar project is a good starting point for this activity; however, it should be recognized that new events could also be relevant, so history is not the only source.

A technique to help focus on specific target areas is offered in a schematic view, such as the *Risk Breakdown Structure* (RBS) (see Figure 9.1).

Schematics of this sort can be used to highlight and even prioritize direction for the identification process. Also, some organizations use extensive

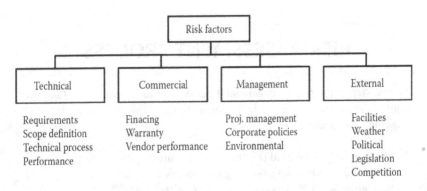

**FIGURE 9.1**    Sample Risk Breakdown Structure (RBS).

checklists to guide the process. The team should not be limited to this set of categories and checklist items but think of these as a guide. One very mature example of this approach comes from an IT organization that created an extensive list of graded questions designed to focus risk assessment on the highest graded areas. This idea is quite similar to a schematic RBS, except this view is in list form. If the list item is for "user support" and graded high as a risk area, that would focus the evaluation process more in that direction. Checklists and RBSs help a risk review group envision how that characteristic might specifically impact future work. Regardless of how the risk items are identified, this step aims to define specific risk events that can be evaluated either at the project level or at a specific Work Package (WP) level. The Risk Register is used to capture event data elements. Each step in the risk model process adds an increased level of detail as the event is analyzed and subsequent additional decisions documented.

A very visible and public example of an unknown/unknown risk impact is found in the 2018 and 2019 Boeing 737 MAX accident scenario. This is a situation where a new and very successful product version was built off of a very mature earlier set of similar versions—each version essentially expanded the original size. There was little reason to suspect control issues with the existing successful pilot navigation software. Yet the new product version suddenly experienced two major crashes, and the root cause was eventually found to be related to a failure of the navigation subsystem. This risk event resulted in significant loss of life and led to a long-term shut down of the entire fleet across the world. Was this an example of an unknown/unknown, or should it have been considered in the test plan? Boeing is a very mature risk-oriented company, yet this problem was not identified prior to product release. The cause of the crash was unknown initially. It took more than 18 months to evaluate and rectify the root cause of this issue. A philosophical point to ponder here is whether Boeing should have been able to test for this situation before the plane was released? It is more important for this discussion of risk to understand that this is not an isolated scenario. In 2011, another similar type of risk event occurred at the Fukushima, Japan, nuclear power plant that was located near the ocean. A tsunami flooded it with extensive damage (IAEA). Was this an unknown/unknown event, or should the location risk potential have been considered in planning and the plant be moved further away from the ocean? Is the potential for a very large wave an unknown event? Both of these events are sobering examples of the impact found in unanticipated risk events.

Each of the example scenarios had somewhat similar characteristics in that the specific event was not handled by the model process outlined here. It does highlight the fact that both of these causal factors should have been identified, evaluated, and resolved. However, it is always easier to second guess this sort of thing after they happen. The two example organizations are mature risk evaluators, yet they both suffered from these unplanned outcomes by failing to identify the resolvable event and mitigate the negative occurrence. Think how much better off both would be if these events had not occurred. It seems clear from these examples that even sophisticated high technology organizations have trouble with risk assessment. The two examples do point out the value of doing better risk assessment. The reader could gain some worthwhile insights into this topic by looking at other public accidents described and asking yourself if these could have been mitigated somehow.

Industry history has many examples of negative risk event outcomes, even in organizations trained in risk assessment. This simply highlights the complexity and immaturity of this management topic. The recommended advice is to do your best, but don't expect to avoid all risk surprises.

The risk assessment cycle time is generally a time-constrained activity, and in all likelihood some identified risks that should have been handled will pop during execution. One can argue whether that should be the case or not, but it seems to be the reality of projects. Recognize that this is a very complex area of concern in that neither the potential events that might occur nor the cost of those events is easily determined. Nevertheless, it is important to follow a model of assessment and handling of such events. A focus on the risk aspect of project management over time will improve the accuracy of the results. Let's move on to the next step and see how the remaining steps are utilized in the overall process.

# QUALITATIVE ANALYSIS OF EVENTS

The third step in the risk model focuses on categorizing identified risks. This helps to prioritize items so that the highest-level risk items will be reviewed first. There are various methods to accomplish this. Typical category groupings use scales of three, five, or ten grades. Unless you have a particular bias or need, we are recommending using High (H), Medium (M), and Low (L) as an initial grading scheme. The logic of the grading scheme is to recognize that this time-constrained process may not allow sufficient time to assess all identified items, so the goal should be to prioritize the list, leaving the lower-level risk items without review. The challenge now is to find a way to display the identified items in such a way as to guide the prioritization. Remember, doing real "productive" work toward finishing the project has priority, and risk assessment is often viewed as a waste of time. This negative bias toward risk assessment will often force the project team to do less of this than they would like, thus leaving more risks without analysis than desired. Given the current level of maturity in the typical project, this is a somewhat reasonable strategy, but it does add a burden to the monitoring step. We certainly don't want the project to fail, but it may be necessary to assume certain risks to save time. This will not be the worst management practice, but one that can certainly be improved on in the future. Figure 9.2 shows a diagram called a *Risk Matrix*.

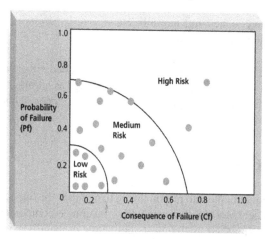

**FIGURE 9.2**    Risk matrix.

The two risk event grade assessments needed to enter data into this format are impact and probability. In this example, each axis in the figure is labeled with a decimal number implying percentage. An alternative grading approach is to assign integer numbers (1–10) or H, M, L grades. With just a little insight, one can see that this is what might be called "fuzzy math" since the values assigned are very subjective, and the grade level groups are somewhat arbitrary. Realize that the role here is to quickly categorize and prioritize the risk events. Note the major domains of the matrix are defined as part of the risk acceptance concept. In other words, we may want to identify and perform a careful analysis of the high-level items, less at the medium level, and cursory review at the lower level. Regardless of the chosen strategy, this is the logic for prioritizing further analysis. The Risk Register will be updated with these grading codes.

We need to add one additional perspective to the risk matrix. Can you see value using ten grade values along each axis rather than three, as shown here? That would give more granularity to the analysis groups, and that could improve the selection mechanics. At the end of the qualification steps, the target categories are chosen, and the following data are recorded:

1. Risk Register reference number
2. WBS ID
3. Description of risk—possibly a short and long version
4. Statement of impact—a code reference could be used here
5. Likelihood of occurrence
6. Coded impact of occurrence
7. Frequency—one-time, monthly, etc.
8. Other items as the process evolves

There is one other data item that will need to be added either in this process step or the next one. That is the event *trigger*, defined as adding some indicator to define that the event has happened or is about to happen. This is an important item in the monitoring phase.

## Quantitative Analysis Step

Various techniques come into play in trying to quantify the impact of a selected item. If the project world was as mature as insurance companies, an actuary would come in with their math tools to define the probabilistic impact of each chosen risk event. This level of quantification would solve much of the subsequent handling problem. A pure quantification process would require the probability of occurrence, dollar impact, and frequency. Using probability and impact data, it is theoretically possible to convert the risk threat impact into an expected dollar value. Doing this for all of the events would generate a mathematical contingency value, and this amount would be set aside to handle the defined future risk events. For example, if there was a 50% probability of an event occurring and the dollar impact for a single occurrence was $10,000, the expected value of this event is calculated as $5,000. The sum of all such calculations would define the project risk reserve required to handle the project risk pool. This math is lovely, but it is not a reality for the project world. The project needs to know how much to set aside to cover risk, but it cannot be obtained through actuarial methods at this maturity point. So, we are left with a "fuzzy math" solution to define how to handle the reserve question.

Rather than a pure math approach, this step is focused on further evaluation of the high-risk items identified in the qualification step. We are now looking at relatively high probability and high impact items. Through this analysis, more details have been collected regarding the event. The Risk Register is again updated, and the candidate set of events moves on the response-handling process, where decisions are made regarding what to do with the event. This process continues until the allocated time is exhausted. Deciding when to abort this step is one of the most difficult in the management life cycle.

# RISK RESPONSE PLANNING

At this stage of the process, the slate of selected risk events has now been defined and prioritized. High-priority items have been further analyzed, and that is now the main focus of the handling process. The key question now is what to do with each of these. Recognize that the same risk event may well be handled differently by various organizations. Different organizations have different *risk tolerances*. Local management should provide guidance on the level of risk they want the project to accept. The worst-case scenario is that some risk event has now been found that violates that threshold. In that case, the item would be presented to management to see if it is sufficient to cancel the project. Absent that, all of the selected events will be handled with one of the following four decisions:

*Avoid.* Changing something in the plan that will eliminate the defined risk. One common example of this group is the decision that using some new technology proposed in the design is not worth the risk, so the plan would be modified to use an older proven version. This might also introduce another new risk, but we will ignore that secondary risk issue here.

*Transfer.* The goal here is to move the risk to a third party who will accept the risk. Several years ago, an author was involved in a product that specified a new metal to decrease weight. Little thought was given to this design decision, and it did not appear on the list to further evaluate. Later, when the first parts were completed and failed in testing, it was found that the metal did not behave like the traditional one. Several attempts to fix the problem uncovered the fact that the internal skill level did not understand how to work with this. Later, a third party was identified, and they had little trouble with it. The problem went away. This was not a perfect example of transferring because we experienced the risk impact, but we were able to transfer it away pretty quickly. This is yet another example to show the role of risk monitoring.

*Mitigate.* Actions taken in this realm are designed to reduce the potential risk's probability and/or impact to a more acceptable level. Once again, let's use a project-level example as a memory device. Assume the project is involved with flammable chemicals. It is decided that an appropriate mitigation strategy would be to have several fire extinguishers available. Would it be worth adding a sprinkler system? What about fire drill training for the team? You get the point! Do something in the right direction and decide when to stop.

*Accept.* Some risks are so small and easily dealt with that it is not economical to spend time developing a formal response mitigation plan. Another example of this is finding a fairly high risk, but low and easily fixed event. In this case, there is no mitigation strategy developed for it. The event is placed on a risk *watchlist* for monitoring in both cases, but nothing else is done to minimize the potential occurrence.

Risk Response planning involves developing one of these responses above to named risk events. The goal of these responses is to enhance opportunities and reduce threats to the project's objectives. Technically, a risk could have positive potential and a negative one, although project teams seldom deal with the opportunity side. We will describe only the threat side of risk here. This activity aims to define an action to be taken for specific risks and the associated steps required to implement this action. It is not difficult to see that these corrective-oriented actions are designed to minimize the negative impact on the existing project plan's scope, time, or cost. Each strategy to alter the current plan should be focused on enhancing project success at minimal additional cost. Once these decisions are made, the project plan needs to be reviewed to see what changes to the current plan need to be made.

Risk events that have been judged in the *avoid* category are particularly meddlesome. This essentially means that some design aspect of the current plan is to be changed. This could have a major impact on the design and the plan parameters. As usual, our goal is to try to give an understandable example. Suppose the original design has specified software utility version 10. After reviewing this as a risk item and finding evidence of questionable reliability, it was decided to remove that version and drop back to version 9. The plan and technical design would need to be modified to fit this. In most cases of this type, changing out one risk item often introduces another risk level, called secondary risk (from version 9 in this case).

The *mitigation* option requires the most administrative and technical work. In this case, the challenge is to find an alternative method to lower the defined risk. This has many alternative dimensions, and it is hard to provide a simpler example than the fire hazard one mentioned above, but this type of risk event occurs frequently—spare tires, seat belts, vaccinations, etc.

Each one of the four handling options affects the risk potential of the project. Recognize that some level of the risk remains, but hopefully, these decisions create a more acceptable level.

Some of the defined risks will require a documented *risk response plan*. Some of these could be very simple if the fix is well defined, while others could be much more detailed. The key idea of a response plan is to make the recovery timely and orderly. If a response action is in the simple to fix category, the documentation would simply name the risk owner and record the item on a *watchlist*. This says that no formal handling is defined, but the *risk owner* is charged to monitor ongoing status. There is one final management action required at this point. All risk event-handling decisions should have a defined *risk owner regardless of category.* This should be an individual who has the best knowledge regarding that item and can quickly lead the recovery steps.

# CONTINGENCY RESERVES

In an ideal case, the decision on the size of a risk contingency reserve would be pure arithmetic, but that is not the case at all. Quantification of probability and impact for the full set is not technically feasible. Also, recognize that the actual size of the risk pool may well be larger than the ones identified so that a math solution would fail even here. This then becomes a messy management question. A risk reserve is needed to handle project risk events and some that are not even yet identified. At the risk of inviting expert criticism on this upcoming solution, let us propose a somewhat non-sophisticated approach. The key question is where to place the reserve and how much of a fund to set aside for this process. Let's describe this in two recommended steps.

The first step is where to allocate the risk funds into a single contingency pool. Many projects scatter these throughout the WBS into various WPs to handle overruns of all kinds. This is the wrong strategy for management. The net effect of this strategy is to pad the whole project unnecessarily. It is important to leave the estimates for the WP as defined in Chapter 6. *Do not pad the estimates for risk in that location.* The goal is to have a single risk reserve to handle risk events as they trigger. When a risk event triggers, the necessary funds will be extracted from the reserve and used to fund the recovery. This accomplishes two things. It takes away unnecessary task padding and better reveals the actual level of risk in the project (even if the reserve is insufficient and has to be increased).

Step two is to decide the size of the reserve fund. That is the million-dollar question today. If there is a history from similar projects, that could provide some guidance. However, if that is not the case, a more general rule of thumb will have to be used. Higher-risk projects would get a larger allocation than lower-risk ones. Given the breadth of project types and related risk profiles, this is an impossible guess (but we will make one anyway). First, try to measure project overruns in the same industry and see if that provides some guidance. One fear is that this will be of little help given poor risk practices. Our rough judgment says that this number is in the range of 10–30% of the project budget, but it could be more for high-risk ventures. The low value would be for low-risk projects previously produced (and there should be some clue from that experience). It is impossible to offer more regarding how to quantify this value. Once the risk assessment process becomes more mature, more data will be available to select an appropriate contingency value.

One real-world management problem with all project contingency data is that it is hard to support the chosen reserve values to senior management. This will require careful discussion to outline the method used and show how such funds will be used. Space does not provide more detail to offer on this topic, but understand that this process needs to be communicated. This topic is the frontier of project management and today suffers from years of flawed approaches and lack of maturing this process. Realize that there is nothing wrong with the somewhat theoretical model outlined here other than the fact that it doesn't yield an auditable result for financial management. It does help to evaluate the risk profile of the project, and that has management value.

# RISK MONITORING AND CONTROL

The risk management problem is not over just because some effort has been taken to evaluate and plan how to handle risks. Risks events will continue to emerge throughout the life cycle, and they need to be handled promptly.

Assuming the planning cycle was accurate, the operational actions for risks are much better defined than would have been the case otherwise. Now, it is possible to link risk events with WP tasks. Also, timely recovery from a triggered event is possible. If funds are required to handle an event with a risk response plan, the risk owner heads that effort and initiates the process of using funds from the risk reserve.

Beyond these more mechanical actions related to handling these events, another cultural approach is for the organization. That is, energizing the organization to be risk aware and proactive. The goal should not be to wait for a risk to occur, but rather be on the lookout for indications. Research studies have shown this strategy to be even more valuable than model-based contingency planning. Motivated humans are great risk evaluators.

## Risk Maps

A sorted view of the project plan can provide insight into when and where to look for a task-linked risk event. This view is called a *Risk Map*. It can be easily created by labeling tasks with some type of risk-level code. For the example shown in Figure 9.3, the code is simply H, M, or L. Note that the H-coded elements show a different sequence view of tasks compared to the classical time-based critical path.

| # | WBS | Task Name | Assessment 1 | Duration | Cost | Total Slack |
|---|-----|-----------|--------------|----------|------|-------------|
| | | Text20: H | | 258.5d | $192,880 | |
| 3 | 1.1.1 | Charter | H | 10.2 days | $8,160 | 0 days |
| 5 | 1.1.3 | Scope Definition | H | 14.5 days | $11,600 | 0 days |
| 8 | 1.1.6 | Budget Definition | H | 14.8 days | $11,840 | 0 days |
| 16 | 1.2.2.1 | Engine Design | H | 81.5 days | $65,200 | 164.5 days |
| 17 | 1.2.2.2 | Mechanics Design | H | 30.8 days | $24,640 | 125.9 days |
| 19 | 1.2.3.1 | Electrical design | H | 30.8 days | $24,640 | 125.9 days |
| 23 | 1.2.4.1 | Body | H | 30.5 days | $24,400 | 0 days |
| 27 | 1.2.5.1 | Chassis | H | 20.2 days | $16,160 | 145.1 days |
| 43 | 1.5.2 | Safety test | H | 7.8 days | $6,240 | 0 days |
| | | Text20: L | | 233.9d | $138,240 | |
| 7 | 1.1.5 | Risk Assessment | L | 5.3 days | $4,240 | 0 days |
| 9 | 1.1.7 | Management Charter Approval | L | 5 days | $4,000 | 0 days |
| 10 | 1.1.8 | Set baseline | L | 5 days | $4,000 | 0 days |
| 13 | 1.2.1.1 | Initial Draft | L | 34.7 days | $27,760 | 0 days |
| 14 | 1.2.1.2 | Integration analysis | L | 10.3 days | $8,240 | 0 days |
| 21 | 1.2.3.3 | Others Elect | L | 27.7 days | $22,160 | 136.4 days |
| 25 | 1.2.4.3 | Design integration revi | L | 10.3 days | $8,240 | 136.4 days |
| 28 | 1.2.5.2 | Other Structures | L | 10.2 days | $8,160 | 0 days |
| 29 | 1.2.5.3 | Mechanics Parts | L | 19.5 days | $15,600 | 140.8 days |
| 30 | 1.2.5.4 | Electric - Electronic Par | L | 11.5 days | $9,200 | 0 days |
| 36 | 1.4.1 | Station Mechanic | L | 6 days | $4,800 | 0 days |
| 37 | 1.4.2 | Station Electric | L | 6 days | $4,800 | 0 days |

**FIGURE 9.3**   Risk map.

Using this alternative format, the focus is on time frames for high-risk grade grouped tasks. Presentations in this format can help activate the watchlist for potential problems. This view is created by simply adding a new data column and providing the risk code for each line item or WP. The view shown is then created simply by sorting on the risk level field.

# SUMMARY

Let us make a few summary points as we leave this section on risk management. First, all projects should go through a risk assessment process as outlined here. Initially, the process will be cumbersome and probably inaccurate; however, each iteration should improve in all steps of the model. Simply getting the team and organization to think about risk in a formal manner will help sensitize the culture to its existence and the associated fact that mitigation strategies do exist to minimize their impact. The last step in the maturation process will be developing techniques to produce a quantification process closer to a true expected value mathematical form. Nevertheless, insight value is gained even through just the assessment process.

Formalized risk management is now becoming widely recognized in both the public and private sectors as an integral facet of effective business practice.

It provides management with a deeper insight and wider perspective regarding effective management of the dynamic project risk environment. Also, risk management at the project level is an essential contributor to success as it focuses attention on issues that potentially affect the achievement of its objectives. Sample advantages from this action are more successful projects, fewer surprises, less waste, improved team motivation, enhanced professionalism and reputation, increased efficiency and effectiveness, and more. Experience suggests that the risk management process does not have to be complicated or time-consuming to be effective. By following a simple, tested, and proven approach, the project team can prepare itself for probabilistic events that may occur. No tool can identify, avoid, or fix all potential risk events; however, the overall risk management offers a process to produce higher success in navigating through this minefield.

*Note: More details on the mechanics of risk management can be found at Carstens and Richardson (2020).*

# REFERENCES

Carstens, D. S., and Richardson, G. 2020. *Project Management Tools and Techniques, 2nd ed.*. Boca Raton, FL: CRC Press.

IAEA. 2011. Fukushima nuclear accident. Available at: https://www.iaea.org/newscenter/focus/fukushima (accessed April 29, 2021).

# Project Communications

# 10

---

## INTRODUCTION

---

According to PMI (2017), project communications consist of planning, managing, and monitoring project communications. The planning portion of communications management includes identifying the best way to provide effective communication to project stakeholders. The management of communications involves the "process of ensuring timely and appropriate collection, creation, distribution, storage, retrieval, management, monitoring, and the ultimate disposition of project information" (PMI, 2017, p. 359). The third aspect of communications is to monitor various project processes that focus on providing stakeholders with their information needs. This chapter will provide an overview of tools to assist the project manager in tracking important aspects of the project to facilitate access to the accurate status of the project and to share information easily.

---

## ROLES AND RESPONSIBILITIES

---

It is instrumental to the project's success that each team member understands where they fit into the project tasks. The roles and responsibilities matrix (RAM) can be used to document which tasks belong to which individuals (PMI, 2017). An example of a RAM is displayed in Table 10.1. The popular

**TABLE 10.1**  Example RAM Chart

| TASKS | RICHIE | ANGEL | NATALIE | MADISON | RYAN | LINDSAY |
|---|---|---|---|---|---|---|
| Develop website | X` | | | X | | X |
| Conduct training | | X | X | | | |

DOI: 10.1201/9781003218982-10

RAM serves multiple purposes as it is an easy way for the project manager to assess which team member may be involved in too many tasks.

Table 10.2 displays a Responsible, Accountable, Consult, and Inform (RACI) formatted chart that clearly defines roles different team members play on the project team based on the coding consisting of who is responsible (R), accountable (A) typically with sign-off authority, the person who can be consulted with knowledge on the task subject matter (C), and the person who needs to be informed of the work (I). This tool is an alternative to the RAM chart. The design difference is just that the RACI breaks out each team member's roles more by displaying not only who may be responsible but also other roles such as A, C, etc., as shown in Table 10.2.

Another way to clarify responsibilities for team members is to provide each member with their roles and responsibilities sheet (see Table 10.3). This roles and responsibilities table is generally shared only between the project manager and the specific team member. This communications artifact helps prevent misunderstandings because it communicates specific tasks assigned to each team member. Another benefit of this sheet is that it can also later serve to evaluate the team member's performance about the progress made within each of the responsibilities assigned. The RAM or RACI charts, along with the individual roles and responsibility sheet, provide direction for each team member and the team as a whole.

# TRACKING TOOLS AND TECHNIQUES

Different status tracking tools and techniques exist to help the project manager and team stay updated on the ongoing status of different versions of documents, commitments on presentations, or demonstrations promised by team members. The commitment-tracking document helps the project manager track formal commitments made by the team members (see Table 10.4). This is a great organization tool that tracks often mismanaged communications promises by logging and tracking them.

There is a communication management plan "developed to ensure that the appropriate messages are communicated to stakeholders in various formats and various means as defined by the communication strategy. Project communications are the products of the planning process, addressed by the communications management plan that defines the collection, creation, dissemination, storage, retrieval, management, tracking, and disposition of these communications artifacts." (PMI 2017, 362). An example of a worksheet tool that can be used to track communication is shown in Table 10.5.

**TABLE 10.2** RACI Matrix

| TASK # | TASKS | COMPLETED | DATE TASK IDENTIFIED | TASK DUE DATE | MATT | BROOKE | JENNIFER | JOSEPH |
|---|---|---|---|---|---|---|---|---|
| 1 | Develop database | X | Month/day /year | Month/day /year | A | I | R | C |

**TABLE 10.3**   Roles and Responsibilities Sheet Example

| CASSIDY BELL'S RESPONSIBILITIES | TASK | SUBTASK |
|---|---|---|
| Task 1 | Serve as lead database designer responsible for assigning work and monitoring work of other team members with a database role | • Identify database tasks<br>• Assign database tasks to team members<br>• Monitor each team members' progress on tasks weekly |
| Task 2 | Serve as lead website developer responsible for assigning work and monitoring work of other team members with a website role | • Identify website tasks<br>• Assign website tasks to team members<br>• Monitor each team members' progress on tasks weekly<br>• Update website information<br>• Design website<br>• Arrange access for all team members to the employee login of the website and shared drive<br>• Arrange access for all team members to the employee login of the shared drive |

An example of a worksheet tool that can be used to track communication is shown in

The Stakeholder Register is another important "project document, including the identification, assessment, and classification of project stakeholders" (PMI, 2017, p. 723). It is a chart that contains specific information on each stakeholder such as contact information, function (sponsor, customer, etc.), expectations, degree of influence on the success of the project, etc. (see Table 10.6). It will be updated throughout the project life cycle as project changes occur.

A stakeholder engagement plan "describes how stakeholders will be engaged through appropriate communication strategies" (PMI, 2017, p. 381). This is "a component of the project management plan that identifies the strategies and actions required to promote productive involvement of stakeholders in project or program decision making and execution" (PMI, 2017, p. 723). This is another input into the management of project communications. An example of a worksheet tool to assist in tracking stakeholder engagement is shown in Table 10.7.

**TABLE 10.4**  Commitment Tracking Document

| COMMITMENT | DATE OF COMMITMENT | TEAM MEMBER MAKING THE COMMITMENT | PERSON TO WHICH THE COMMITMENT WAS MADE | DATE COMMITMENT IS TO BE COMPLETED | COMMENTS |
|---|---|---|---|---|---|
| Website demonstration | Month/day/ year | Project team member Mark Bell | Customer Stan Thomas | Month/day/ year | Team member Cassidy Jones will present the demonstration |

**TABLE 10.5** Communication Management Plan

| STAKEHOLDER | TYPE OF COMMUNICATION | SCHEDULE | RESPONSIBILITY | COMMUNICATION FORMAT |
|---|---|---|---|---|
| Project team | Weekly status review | Every Thursday at 10 a.m. | Project manager | Meeting/teleconference |
| Program manager | Weekly team progress report | Close of business on Monday | Project manager | Email |
| Program manager | Monthly team progress report | First Monday of every month at 10 a.m. | Project manager | Meeting |
| Project sponsor | Project review | First Tuesday of every month at 10 a.m. | Project manager | Meeting |
| Customer | Customer reviews | Second Monday of every month at 2 p.m. | Program manager Project manager | Meeting with the customer and responsible team members |

**TABLE 10.6**  Stakeholder Register

| NAME/ POSITION | CONTACT INFORMATION | TEAM ROLE | DEGREE OF INFLUENCE | DEGREE OF SUPPORT | EXPECTATIONS | INTERESTS |
|---|---|---|---|---|---|---|
| Pam Carre/ Program Manager | Phone Email Address | Sponsor | High | Medium | Monthly communication, status updates on customer satisfaction with deliverables | Enrichment and advancement of project team members |

**TABLE 10.7**    Stakeholder Engagement Plan

| STAKEHOLDER ORGANIZATION, GROUP, OR INDIVIDUAL | PRIORITY | COMMUNICATION STRATEGY | PROJECT PHASE | ENGAGEMENT TOOLS |
|---|---|---|---|---|
| Risk manager | Medium | Engage during project kickoff through project close-out | All | Face-to-face Email |

The Issues Log displayed in Table 10.8 tracks the continual flow of new issues raised throughout the project life cycle (PMI, 2017). The example shown here has many different columns to track the issue number, report date, description of the issue, and name of the person that reported the issue. It also lists the degree of urgency to resolve the issue, assigned team members responsible for the resolution of the issue, status of the issue, solution to resolve the issue, and the close date that represents when the issue was resolved.

# SUMMARY

This chapter addresses the value of various tools that assist the project manager in providing effective communication with different stakeholders. Different roles and responsibilities tables were discussed as a means for the project manager and team members to document individual task assignments and which team members work together on specific tasks. Tracking tools such as the commitment-tracking document and communication management plan were discussed to help a project manager and project team track the frequency of team activities such as meetings, work assignments, commitments, and documentation control. A Stakeholder Register and Stakeholder Engagement Plan was discussed that represents a tool for the project manager to track important demographic information on the project stakeholders and the best way to engage these stakeholders during the project. An Issue Log is another tool useful in tracking ongoing issues as they emerge throughout the project life cycle. This tool is useful in assigning a responsible person to follow the resolution of an issue. Without effective communication, a project will face many challenges and possibly fail. The effectiveness of communication skills tends to make or break a project managers' longevity in their field. Also, failure to communicate is often ranked as the number one reason for projects to fail.

**TABLE 10.8** Issues Log

| ISSUE LOG # | REPORT DATE | ISSUE DESCRIPTION | REPORTED BY | URGENCY | ASSIGNED TEAM MEMBER | STATUS | SOLUTION | CLOSE DATE |
|---|---|---|---|---|---|---|---|---|
| 1 | Month/ day/ year | Broken link on website | Alec Monroe | High | Craig Mellon | Closed | Link fixed | Month/ day/ year |

# REFERENCE

PMI (Project Management Institute). 2017. *A Guide to the Project Management Body of Knowledge*, 6th ed. Newtown Square, PA: PMI.

# Speeding Up the Project

# 11

## INTRODUCTION

This chapter describes various methods to speed up the completion of Work Packages and the project itself. Realize that each technique to decrease time has some reactive countermeasure (i.e., increasing cost, increasing risk, decreasing quality, etc.). The assumption made here is that the original plan was considered optimum in producing the desired product. Anything done to decrease that time would be considered less than optimum. That said, there is no phrase heard by the project team more than "why is this taking so long?" In reaction to this pressure, the team will be looking for strategies to cut cycle times. In some cases, the selected speed-up strategy may result in a longer time. The techniques outlined here do work as advertised, but be careful of the hidden backlash that may result.

Each of the techniques described here needs to be clearly understood to effectively utilize the technique. One of the prerequisites of speed is to have a well-thought-out project plan, yet even the planning process itself is often challenged as taking excessive time for no value. As a key strategy, there will be little argument in the belief that having full-time future users in the project team will help speed up the result. Similarly, having an engaged project sponsor can help cut out bureaucracy. Each of these ideas is often lacking, yet the pressure by these groups often remains. One interesting test of how serious the stakeholders are would be to suggest that all changes will be deferred until a later phase. Rejection of this idea suggests that they have multiple goals.

How one views the question of project speed is often based on their belief structure. For example, is planning worth the time? How about risk assessment? How much overtime is ok? How much time do you spend on stakeholder communications? So, the choices for speeding up the project lie in this convoluted domain of belief options. With this introductory warning regarding strategies to solve the "need for speed," some sample major techniques will be summarized below.

DOI: 10.1201/9781003218982-11

## Planning

The speed bias conflict here is that planning is inaccurate, so why do so much of it? With this logic, there is a somewhat prevalent attitude that less should be done early, thereby saving time. Admittedly, if the team is not focused on this negative planning attitude, there is a tendency to spend more time planning than necessary. Some call this *Analysis Paralysis*. The popularity of the *agile* model (Chapter 15) l is likely a result of that opinion. Completely cutting out any of the life cycle sections outlined in this text is a mistake, in our humble opinion. That said, there are ways to use templates, historical data, and other shortcuts that can speed up each of the formal processes outlined here. So, the first rule of thumb is to do the process as described, at least to some minimal level, but use various tools to speed up the cycle. Remember, success also is measured by budget, product functionality, and risk parameters, as well as schedule. Surprisingly, if one were to add user satisfaction to the project outcome goal, the actual life cycle might take longer. This also would be counter to cutting all change requests to decrease time. Recognize that a singular speed increase goal may be naïve.

## Adaptive Life Cycle Models

Since the 1980s, there has been a movement to change the waterfall model of planned sequential steps into a more fluid process with less up-front planning and more active user involvement. This is called the iterative model, where the end goal is now defined as when the user is satisfied with the result (or the project ran out of money). There is certainly merit in the idea of active user involvement as the modern iterative model dictates. Too often, the team is left isolated to interpret requirements only to find out later that they did not understand. Regardless of the life cycle approach chosen, a major universal key to speeding up the project is through active user involvement.

During the search for a better method of executing a project, the following characteristic emerges as causing the traditionally planned project to be less successful:

- First of its kind
- Requiring creative solutions that cannot be defined in advance
- More of an exploratory need without a clear definition of the goal (i.e., developing a cancer drug)
- Resource needs are unpredictable or not in good supply

In these environments, there is more of a need for iteration with less up-front planning. In 2001, a group of software practitioners met to summarize this approach formally. From this meeting, they created a *Manifesto* of project practice called *agile* (or several other dialect names such as *Scrum*). The original target for this new method was software projects, but more recently, the targets have expanded into other project-type areas.

There is now an industry debate regarding how well agile covers some of the needed life cycle processes. It is best to leave this debate until later in the text when you are more prepared to see both sides of this question. Chapters 15 and 16 will open up this question for more extensive review. Regardless of one's position, there is no argument that the agile-oriented minimal up-front planning approach to project management is growing in usage. The *need for speed* is surely in focus here.

# RESOURCE MANAGEMENT

Previous discussions have highlighted the need to track the availability of resources needed for task execution carefully. *This is the number one recommended speed-up strategy.* If one wants to explore this idea at a more global level, it is recommended to see details regarding the Critical Chain (CC) model. Richardson and Jackson (2019) offer a detailed discussion of this methodology format based on the application of Goldratt's (1997) *Theory of Constraints (TOC)*. Management of the CC elements is handled using resource alerts and buffer management for the defined task chains. Implementing these concepts will require a cultural change throughout the organization, beginning with a significant shift of focus from task control more to buffer management. Space constraints here do not allow more description of this somewhat complex model, but the authors' opinion is that the future of project management will move toward this concept. Essential elements of CC* are the elimination of task padding, using buffers to control overruns, and carefully managing resource availability. Note that many of these ideas have been mentioned throughout this text. Even if you choose not to do a radical redesign of the traditional approach, the concepts of this model are worth exploring.

## Fast-Tracking and Crashing

The traditional project world has long recognized the terms *crashing* and *fast-tracking*. Regardless of the project methodology employed, these concepts are

---

* See more CC background at https://goldratt.co.uk/ and other sources

valuable tools for the project team to use. The following is a quick overview of these two approaches:

> *Fast-tracking.* This technique moves two sequential tasks into a more parallel relationship. Time can be saved if some part of this set can be overlapped in time. For example, if a five-day task can be overlapped by two days, the set cycle time is decreased by that amount. For this method to work as indicated, the necessary resources must be ready to start as defined.
>
> *Crashing.* The best mental model for understanding crashing goes back to the "digging a hole" estimating example introduced in Chapter 7. If it takes 80 hours for one person to dig the hole, two people should be able to do it in less time. This technique trades resources for time, and that is a crashing decision. This is not pure arithmetic as it may not be feasible to allocate 80 workers to finish the hole digging job in one hour. There is a limit to the increased allocation logic, but the idea is a sound and useful tool. Once again, we see that proper management of the project resource is a key factor in timely execution. This is related specifically to the quantity and skill of the project team.

Both of these techniques are considered classic methods of time compression and should be used as appropriate.

## Working Overtime

There is hesitancy to mention this technique in many ways because it is often overused and may not be a true time reduction method. In concept, this is a subtle method of crashing. If the plan assumes that a resource is working 40 hours a week but instead is working 60 hours a week, the elapsed time to execute a task has been theoretically crashed by 50%. Time, in this case, has been decreased, but the planned cost has been increased by the overtime rate. Of course, if the resource is not paid overtime, free time has just stolen time from the individual. There is also the recognition that productivity likely declines if this method is used too long. Be careful of decreased morale and lower team productivity from what appears to be a free crashing strategy.

## Resource Control

A second resource-related speed technique involves the timely allocation of resources to tasks. As trivial as that statement sounds, it is one of the biggest time wasters in the life cycle. If a resource is needed to perform a task on Monday but does not arrive until Wednesday, two days of schedule are lost, and the time is not recoverable. Something as simple as having timely resources in place is frequently not practiced.

# MISCELLANEOUS TECHNIQUES

The items mentioned above represent specific operational methods related to speed up completion. The list below represents a more general management process that is relevant across the entire life cycle.

*Communications.* One of the recognized areas of project management that is central to success is good communication. This includes links between the internal team, senior management, and stakeholders involved in the project. Project managers who are not so inclined to communicate will suffer in this area. This may well be the most common shortcoming of the management role. When individuals do not know what to do, time is wasted. The term *coordination* possibly fits best here.

*Scope Changes.* This topic and the role that zero changes could make in the life cycle have already been mentioned, but they are a fundamental root cause for schedule overruns. Changes are disruptive to work productivity, and they drain resources away from doing approved work. If the true goal is to cut costs, then freeze the scope! That will measure the priority of this goal with your stakeholders, yet in most cases, it will be rejected by them. Probably tight control of changes will be the best compromise available here.

*Risk Assessment.* If no risk assessment has been done, every risk event that emerges becomes an unplanned problem to resolve. Some of these could have been identified earlier and avoided or mitigated, as Chapter 9 has previously outlined. Some might believe that cutting out the risk assessment process will speed up the completion time, but think about what happens when an unidentified major risk emerges that could have been dealt with during the planning phase and then reassess this idea. To speed up the overall life cycle, some level of risk assessment is needed, and the organizational risk culture maturity is also important as a speed-up factor.

*Managing Stakeholder Wants and Expectations.* This aspect of project management is viewed as the new kid on the block. The project stakeholder community offers both opinions and influence on the project. In some ways, this is part of the communication gap issue mentioned above. As a personal example of this, one of the authors was working on a large project where an undefined risk issue emerged that required a long lead time expensive hardware item to resolve. This normally required many months of procurement-type justification to get approved. The project owner happened to be the organizational budget manager who approved such requests. Surprisingly, the request was resolved in days with minimal paperwork, and this major problem evaporated. Even though this is a radical example, the value of having supportive stakeholders cannot be underestimated. Without the involved stakeholder mentioned above, this project might have failed on the spot. Other complex decisions were also later resolved with the sponsor's help, and a very complex project was finished successfully. Having active stakeholders involved in your project is invaluable.

*Team Morale.* Just like any team, the morale of the group can be pivotal to success. Relationships within and external to the team are important parts of wading through the daily dynamics. It is hard to remember how often we have had to go to someone in the team and say, "I need some help" or "I need a favor." Without good morale and reasonable relationships, this type of interchange won't work.

*Competent Team Leader.* Even though productive teams do not need much management oversight, it is important for management to react quickly when they do. The problem with this statement is how to define what that means. The project manager needs to be competent in leading highly technical individuals who have both above-average intelligence and egos in most cases. In a classic view of the successful leader, the team needs to trust the leader more than lean on him for guidance. Leadership results suggest that the role of a project leader is best served in initially communicating the project goals, then letting the team evolve without tight controls into a more self-directed model. The key point here is that the project manager is often not the best person to decide how to do something. If the worker is properly trained, he might well be the best source. In any case, management's primary role should be more in orchestrating the team to become focused on the project goals. Learning to accomplish this is a lifetime challenge well beyond what can be described here. A competent project manager will be able to execute the various processes outlined in this text and will have the fortitude to persevere through daily challenges. If one can do that, one will be successful in many organizational roles.

We'll close this section with one controversial statement. Does the project manager have to be a technical expert to be successful? Our opinion is no. Recognize that there are two dimensions to this role, technical and managerial. Being technical helps with knowing which direction to go when faced with complex situations, while being managerial deals more with the human communication and operational coordination process aspects. Having an excellent technical team lessens the need for the PM to be technical; however, there is the need for some balance.

# SUMMARY

The focus of this chapter has been on various concepts and approaches to speed up the project. That theme will remain regardless of the life cycle methodology selected. Any shortcut taken probably has some negative aspects related to it. The challenge is to recognize that speed must be a goal properly evaluated. This decision must be made in collaboration with senior management and stakeholders. The team will tend to be much more conservative regarding time compassion than outsiders would like. Individuals pressing for a quicker time should be offered an internal role in the team so that they can become more sensitive to the trade-offs. There is also a danger here that the new team member cannot understand the environment and will negatively impact the team.

# REFERENCES

Goldratt, E. M. 1997. *Critical Chain*. Great Barrington: The North River Press.
Richardson, G. L., and Jackson, B. M. 2019. *Project Management Theory and Practice*, 3rd ed. New York: CRC Press.

# Monitoring and Control Processes

**12**

---

## INTRODUCTION

---

When project operational and goal variances emerge, the management goal is to move the status parameters back toward target goals. This is the basic role of monitoring and control. The key concept of control says that you can't accomplish this role unless a measurable target is established. This involves defining the plan baseline, and Key Performance Indicators (KPIs) are used to signal corrective action. This chapter outlines the management threads that constitute the basic steps for this activity.

The planning process is a vital part of project success, but invariably operational and goal variances occur later. Making timely decisions that guide the plan back toward its target goals is the role of monitoring and control. Many organizations view this process as simply comparing current plan values to actual values. Admittedly, that is one view of the process, but it is not the main theme for a more mature approach. The concept of goal baselines has been mentioned previously, but in this stage of the life cycle, their role becomes more visible. Some projects view a plan as the "goal of the month." That means each month, the plan is reset to new targets based on task slippage. In this form, the plan is more of current status or wish statement with little real value for a control. This statement means that the control process should work in concert with the baseline concept, not a monthly changing one.

Improper use of certain status variables can be both non-productive and misleading. Also, recognize that measuring the status of a baseline parameter does not cause something to happen. It merely indicates a particular status, and it will still require human intellect to decide what to do about it. For the baseline concept to work as a control guide, defining some status parameter that evaluates that baseline is necessary. These parameters will need to be defined for the various approved outputs, inputs, or other targets. It is common practice

DOI: 10.1201/9781003218982-12

in some organizations to use the term baseline to mean "our current estimate." An argument against this is, "does it make sense to compare the current status to a monthly updated plan?" Our answer to this point is, no! The original baseline should be kept as approved, but senior management can set another baseline for comparison purposes based on some unique situation. As an example, modifying a baseline because of weather-related delays may make sense. This external event is beyond the project's control, and in this case, setting a second baseline to show the impact of bad weather is logical. If baselines are meant to be targets, it does not make sense to punish the team for bad weather. We will assume all baselines are fixed for the duration and all status measures will be compared to these fixed values for this discussion. Baseline practices vary considerably, but the preferred model is to keep them as pure as possible so that they can compare progress to an initial approved target.

Many things can go wrong during project execution, and it is difficult to list only a few examples. Let's take a more global philosophical approach by using Murphy's Law as our guide. Based on this, look at the project outlook status view as "Anything can go wrong, and it will at the worst possible time." That is the best guidance for thinking about the control process. This is where well-defined status measures provide insight and guidance into resolving the problem. Beyond this universal vision approach, some specific areas should be highlighted, but a more formal control theory needs to be defined before describing it more.

# CONTROL MEASURES

Much of the control process involves counting things (i.e., hours, days, dollars, people, etc.). What is missing from this view is the context into which we place these counts. Let's define this data group as *metrics*. A working definition for the term is "a measured property of a process or product whose possible values are numbers or grade values. *A measure is a specific value of a metric*" (Parth and Gumma, 2003). For this to be used in the control process, the role of various status metrics must be better defined. A second term, *Key Performance Indicator*, is used to add context to a particular measured value. Named KPIs help guide the project toward intended goals. The adage "What gets measured, gets done" summarizes how the use of metrics and KPIs can help identify problem areas and guide corrective action. Realize that a KPI value does not cause action, but only points to particular areas from which a response may be needed. A project typically measures and compares its overall performance by collecting comparative data for defined KPIs.

A comparison is made to a related baseline value to add further meaning to a raw KPI value. When a KPI is defined, it is intended to indicate some meaningful comparative performance of the project. For example, if we define a KPI as the Estimated Completion Date, supplying a value to that is designed to convey a current schedule status meaning. One can see that it would not take much time to create 30 or more of these to portray or track various status parameters. But there is more complexity to this process than just creating and tracking 30 KPIs. Let's look a little deeper at some of the characteristics of a KPI. The BrightGuage web blog offers a list of 11 types of KPIs (BrightGuage):

1. Quantitative—measured by a number
2. Qualitative—not measured in numbers
3. Leading—used to predict future outcomes
4. Lagging—measures what happened (accounting data)
5. Input—measures items coming into a process
6. Process—gauge efficiency of a process
7. Output—measures output (i.e., volume, profit)
8. Practical—unique to specific situations
9. Directional—evaluation; trends (how long to fix)
10. Actionable—measure company commitment; culture, discrimination

The list above represents both an answer and a problem. The use of too many KPIs can be overwhelming. There is both a collection cost and an understanding absorption issue related to each. The challenge is to select the fewest KPIs that will best reflect the primary goals of the project. *Case study research suggests that the number used should be in the five to ten maximum range.* There are differing schools of thought regarding standard names for KPIs. That list should be custom-created from the control needs of a particular project. One rule of thumb is to select KPIs that point out areas where the project is drifting from approved baselines or showing a trend that needs timely corrective action.

One of the status metrics that will be visible in almost every project is the target values for schedule and budget at completion. During the execution cycle, various schedule and financial status data will be presented to stakeholders. Be wary that raw data of this type can be deceptive unless it truly compares current status to planned work. A typical misuse of such data comes from the typical presentation that compares planned versus actual budget status. This comparison may show the project having spent 10% less than planned—great news, everyone says. This is a very typical misuse of a KPI. Just because the project has consumed 10% less budget than the plan, is the project doing well? The more basic question is, "What has been accomplished during this time?" A more accurate comparison can be determined by using a technique called

*Earned Value* (EV) which adds a third measure indicating the amount of work completed. Remember, measuring resources consumed has nothing to do with what was accomplished with those resources. The warning here is to be careful of KPIs that purport to show simple comparisons of a single variable pair, such as plan versus actual. Be aware that this status misuse syndrome is not the only possible metric distortion in common use.

A KPI should also be used to help the team manage the project. In simplistic form, the following two questions should be supported by KPIs:

- How is the project doing compared to approved baseline values (Gap data)?
- Where are the specific process thread variations, or what are the root causes of those?

Simply stating some metric accounting-type value that happened in the past is useless without context. Appropriate measures should help to define future outcomes. Each has its nuances, so no one KPI group can claim that it is the silver bullet or magic formula that will fit every monitor and control requirement. We will stop this basic theory discussion at this point and look at more specifics on the use of KPIs.

Having a plethora of project status data is comforting, but one has to ponder the question of collection cost versus relevance and decision-making value. This is yet another area where modern software systems aid the project management activity by capturing various status data as a by-product of the built-in tracking mechanics. For example, Microsoft Project collects hundreds of raw status measures as part of its calculation processes, so the cost of collection is minimal; however, that still leaves the question of interpretative value. Be aware that manual data collection can be expensive, but a comparative warning is that automatically generated computer data can be overwhelming in volume and misrepresent actual status. If a data element is not helpful in decision-making, is it worth collecting?

One point that has not been mentioned until now is the status presentation format. There is a growing affinity for graphics and color to be added to the data. Traditionally, red values indicate below plan, yellow means the measure is in the warning area, and green indicates the value is within plan tolerances. It is easier to identify gap target areas in this manner.

The project environment has been characterized as being very dynamic regarding status variables. KPIs detect where variations exist and hopefully help point to the root cause. Also, recognize that the management side of the control process is not singularly focused on data. Observation and verbal communication, along with other sources both inside and external to the project, are vital pieces of the management puzzle.

Collectively, these defined process threads represent the management control focus areas. One can see by the breath of the control areas described why project management can be a hectic daily environment. One physical metaphor to help explain the management side of this process is to envision a set of project control knobs representing decisions the project manager can make to resolve gap issues. For example, if resources additions are needed, the resource knob can be turned to simulate adding more. Likewise, if more budget reserves are needed, this can be simulated by turning the budget knob. Of course, the real challenge is to have appropriate supporting data that helps identify the exact cause of the variance. A project manager's demonstrated skill in determining the root cause of performance gaps will determine his success in the long term. Developing an appropriate suite of KPIs is a major contributor to this activity. Without data to accompany a decision, one only has an opinion to guide them. "I think" is not the recommended KPI. To effectively utilize KPI data, it is necessary to understand the associated control thread processes. Hopefully, the brief explanations described here will provide sufficient insight into the essential management logic for monitoring and status. The sections that follow describe the major process threads that need to be dealt with to support the project-approved goals.

There is another somewhat ignored management issue that needs to be mentioned. In many cases, typically all negative events related to the project are viewed as the fault of the project team and its manager. It is important to protect the integrity of the team in such situations where the root cause lies elsewhere. KPIs should also be used to help point to external sources that are the root cause of such variations. If the problem cause is internal to the team performance, then so be it. However, if the root cause lies elsewhere, an effort should be made to gracefully point to that source as well. This is a conflict source and should be approved carefully with well-defined data. Exposure of below plan external support is a tough personal trait activity for many managers, but it is important to maintain the reputation of the team. All conflicts should be embraced positively and professionally. Control data will be a major support to deal effectively with these conflicts. These actions should not be finger-pointing but "analyzing" the problem.

# Earned Value (EV)

EV, as previously mentioned, is a sophisticated project status metric growing in usage. This metric deals with introducing a concept called "earned," in which measured task accomplishment is compared with planned and actual resource consumption. This three-variable metric overcomes the traditional process error of simply comparing plan versus actual values. This is the only tool in the group that can forecast project completion based on trend data. A detailed description of this KPI metric calculation can be found in Richardson and Jackson (2019) or Carstens and Richardson (2020).

## Schedule Control

Schedule variance is one of the most visible gap control focus topics. A measured schedule variation causes a further search for the root cause and, in this view, it is only the tip of the status iceberg. As an example, think of a schedule status gap as the result of other issues. Here is a common example. If the quantity or proper skill resources are not applied according to plan, the schedule is likely to slip. Once a variance is identified, the challenge is to trace backward and seek out the reason for this. Chapters 7 and 8 have shown the proper mechanics to build the plan and set an approved baseline. Using this base data, the cause of the variability can be traced back to the WBS segment and WP detail. Failure to follow an integrated process as described limits the ability to effectively trace such variances. This is yet another example showing how a clean linkage from scope through execution has management value.

A significant schedule control view comes from the critical path and the risk map definitions defined in earlier chapters. The critical path is a schedule roadmap for time focus, and it helps track specific critical tasks that will most affect the schedule at any point in time. If there is one element of schedule data that is most used, it should be the critical path. A second and little-used view of potential task variation data is related to the timing of risk events. If a reasonable risk assessment has not been done, this view will not be available until the actual event occurs unexpectedly. Risk management artifacts constitute a valuable analysis tool that is present in high maturity organizations but little seen in less mature ones. Similarly, process gaps in other control areas often create variances in other output variables, such as schedule or budget. This crossover effect is common across all of the process areas.

## Budget Control

Because of mature accounting systems, there is no other process area where so much status data is collected. The accounting profession has matured for hundreds of years and will have extensive systems for producing this class of data. Unfortunately, without careful integration between the financial side and the project plan, this raw data may be of little value to the project. The previous discussion described how Work Breakdown Structure (WBS) codes need to be attached to the organizational work collection data to map the two together.

Otherwise, there is no integrity in data comparisons across these two systems. Chapter 8 described the desired internal resource environment using these codes. This previous section also showed how Control Accounts (CAs) should be established as either a single WP or a defined grouping of these. Through all of this structuring, the goal is to help define elements of the project that would link outcomes to other data types. Once again, failure to structure the project plan properly sabotages the future value of data collected. If this point is not clear, review the budgeting chapter material related to WBS codes and CAs. A large part of the overall control strategy comes from this type of data discipline so that any status data collected can be linked to the project structure through WBS codes. This needed linkage seems obvious, but it is not a well-practiced concept.

## Product Control

A third common control variable for the project is some measure of product status. This is typically a quantity or quality measure for the product. In many projects, the focus is only on designing the product, and tracking the evolving actual product performance status is not the focus. However, contemporary quality management systems in some organizations are more focused on how the product is evolving through the life cycle. Designing a product that doesn't fit the approved baseline is a failure by definition, no matter the status of other variables. There are cases where project supply expenditures were on plan, but the product was not progressing well at all, and much later, the project was canceled because the product was unacceptable. It is hard for project teams to admit that the product is not going to be successful, and they believe that for too long until it becomes clear that the project must be canceled. Part of the mantra of planning and control is to also cancel doomed efforts promptly. Conserving organizational resources by canceling a failed project is just as goal-appropriate as managing the approved resource status for a successful project. The control process should support both ends.

A common behavioral characteristic among project teams is the project becoming their baby, and their professional ego gets involved. One does not want to have their baby called ugly, nor does one want to see their baby killed. That simple statement explains a significant project control problem. There is a tendency to hide control variances from senior management decision-makers and believe that the variance issue can be fixed. The project manager's role is to raise the healthiest baby possible but also to inform the project owner and the stakeholders with an honest assessment status of what is happening. The intent here is not to be morbid with the baby analogy but to understand that "status honesty" is a requirement. It is better to inform management of negative projections and related corrective action plans now than have a negative surprise later. This statement is true of all of the outcome-oriented status areas. An honest status of the project must be a timely shared operation, and all parties should work to achieve the outcome possible; even failure status should be shared.

# Scope Control

Scope control is the one control thread with a well-defined management control process, meaning the goal is clear and the tracking process is defined. Scope control is a universal problem, and it is one of the most impactful areas affecting the outcome. The good news here is that it is manageable if the impact is understood, and a proper control process has been approved by management. Here is a simple scenario example. A senior stakeholder asks for changes to be made to the scope baseline (i.e., new work required). The team wanting to be supportive performs this work without asking for approval. This process repeats over and over again. No recognition is given to the related schedule or budget impact. This project status metrics show a failure by definition, and the team performance would be judged ineffective. Managing scope requests must be done formally, and this is a common source of control mismanagement. Recognize that scope changes also change the original basis that the project plan was created for.

The formal process name for scope control is the *Integrated Change Control* (ICC) system. Failure to have such a system sabotages the concept of project control, and a disaster awaits. The first organizational scope control step is to establish an ICC management board function that includes both technical and business focus (typically two or three members). These individuals need to be familiar with the project's approved scope and understand the "big picture." They are organizationally aligned with the project owner and have delegated authority to approve scope changes within the specified levels. All requests for scope change are presented to them (let's assume all requests). The board has the delegated authority to accept or reject each request. No organizational entity should have the authority to bypass this process (do you see a possible issue here?). To legitimize this function, it is important to have a formally signed statement by appropriate senior management mandating that this will be the only way to change the scope. This important control body stands to defend against runaway project scope. This is one of the most important control steps in the project management structure.

Unfortunately, there is an associated management layer to the scope control process that is quite often omitted. While it is good to keep control of scope changes, the related question is what should happen when a change is approved? Doesn't it make sense to suggest that if a scope change adds work to the project, the approved plan should be revised to reflect that change (i.e., schedule, budget, etc.)? Clearly yes. Without this action, look what just happened to our wonderful approved baselines for schedule and budget! For reasons that are hard to fathom, in most cases, there is no baseline adjustment made or even recognized for this situation, and the approved baseline schedule and budget remain unchanged. This somewhat hidden growth in approved scope will destroy effective schedule and budget control. Unfortunately, there is no easy way to fix this unless the local management culture can be modified. Logically, if the scope increases by 25%, there needs to be a way to add compensating changes to the plan. To resolve this obvious shortcoming, the project team has two bad options. The first is to pad tasks sufficient to cover this variability, or a second option is to just let the project appear to overrun the baseline. Given the negative perception of overruns, it is not hard to guess which path is typically chosen (tsk padding). The typical band-aid fix for this is to add schedule and budget padding and use these artificial uplifted values to cover variances caused by the approved scope changes. That action hides a true status and destroys the integrity of the "approved" project baseline. From a management viewpoint, WP estimates should represent the honest estimates for planned work and not some randomly added padding to cover yet unknown scope change. The proper goal should be to measure the actual work status against the original plan. This scenario is the Achilles' heel of scope control.

The scope growth problem is clear, so now let's describe an approach for managing it. First, it is important to point out that most organizations neither freeze baselines nor recognize scope change measurement. The scope problem that needs to be solved is how to keep integrity in the original plan definition as ongoing scope-enhanced status is measured. From a theory viewpoint, when a unit of work is added to a project, some recognition of that should be added to the approved plan. Also, the baseline should be changed to something like "current baseline," which is a measure of added scope. This theoretical approach is valid, except it is difficult if not impossible to implement. Some of the scope changes are very small, and it would be hard to define their impact, while others could be reasonably estimated. A middle ground to this process is to use pooled reserves to keep this segment of change separate from the direct plan and show approved change levels inside the scope reserve pool. The budget portion would be easier to set aside, but the schedule reserve for change would be a little more arbitrary. It may not be practical to change a WP to show the new scope, and it also may not be administratively productive to do so, but the one thing to say is that it is not fair to the project team to be held to the original baseline metrics while the ICC process is adding 25% to the defined work. Team status should be measured using the actual project scope as it expanded, and not blame the team for an apparent overrun resulting from approved changes. A scope-oriented KPI can help with measuring this phenomenon. Managing this control thread in a real project environment is the tougher question. As with many good management ideas, this one has a valid definition but is difficult to implement. Be sensitive to the issue defined here and segment the scope change data from the original baseline plan.

# Resource Control

Resources represent the process engine that drives the project. This can mean human resources, material, facilities, or dollars. The absence of timely availability to any segment of this group will slow the engine down. For this reason, KPIs related to resource supply are common.

A key resource-tracking parameter should monitor the quantity of these factors and the timing at the unit WP level, especially the critical path units. Assume for a moment that the work estimate for a critical path task is correct. In that case, the related action to complete it on time is to have the proper quantity and quality of defined resources available at the needed starting time. External resources should be in tight partnership between the project and the resource owner. For this group, the project plan needs to be considered a "contractual" requirement statement for this delivery. Various status KPIs can be used to trigger the needed schedule for resources. Lack of resource control for the project is often the number one reason for schedule overruns. Failure to control this process area leaves no easy way to know whether the overrun was caused by poor productivity or just not having the resource in place. It should be clear that not having a resource for a WP means that nothing related will be performed in that period.

# Risk Control

Chapter 9 described the risk-monitoring process. We will reiterate here that failure to recognize risk as a threat to project success is short-sighted. It has been stated that the current level of sophistication regarding risk management is immature in most project environments. The alternative to a formal risk management process is the John Wayne school of risk management (i.e., when the bad guy shows up, just get a big gun and shoot it). This alternative strategy means that the control logic is based totally on the monitoring side. The previous statement regarding the scope management task as being a poor practice also holds for risk. The management rule here is that a formal risk assessment process will lead to the establishment of a risk reserve, and all risk-related actions should use that contingency reserve as needed.

## Communications Control

The term communications control goes against the theory of one access. Research studies indicate that ineffective communication is the primary root cause of project failure. The key issue here is honest and timely communication. The PM is charged with keeping these various channels open regardless of the message. As an example, filtering project status to hide variances is one of the primary examples where poor communications practice hurts. It is important to recognize that erroneous communication is as bad as no communication. The goal for this area is to make sure that all stakeholders have access to appropriate project information and, to support this goal, develop a formally defined method where all participants have one formal place to obtain this access. Chapter 10 has described more details on various related processes.

## Issue Control

A point made multiple times is that the world of projects is dynamic and driven by Murphy's Law. The process areas outlined above have described where much of the variability occurs; however, there is yet one other control area that seldom gets discussed. We will title this area as "issues." The tough part of this is to define an issue. A somewhat loose *definition is any project event or activity that is not part of a planned WP activity and not working correctly.* Here are some examples:

> The copy machine is broken, and the vendor does not have the part to fix it quickly (i.e., quick is the maximum amount of time the team can reasonably be without a working copy machine).
> An employee is sick and will be out for a month, but a replacement is needed.
> Management needs some special data collected.
> A senior stakeholder has requested some unique status data that will take a few hours to obtain.

Many items in this category emerge daily. The normal strategy is to pass the problem to a team member to handle. Note that none of these sample issues were shown on the project plan and, in many cases, may not be listed in a person's basic job description; however, each issue needs to be resolved in a defined time frame for smooth operations. The inherent hazard here is that the item will get lost in the fray and emerge later to become a more significant problem. So, the operational goal for an issue is to ensure that each is handled as needed by the defined individual (and not get lost). The assumption here is that the item is fixable in a reasonably defined period; otherwise, it needs to be elevated to something like a risk event, which is a close cousin to an issue. Some will argue that an issue and a risk event are the same, but we choose to keep them separate so long as the defined issue can be resolved in a normal cycle time. When the issue process is left as verbal communication among the project team, there is an increased chance that it will get lost in the daily worklist. Based on this belief, the best approach for handling this oft-ignored control group is to formalize an *issue-tracking system*. Documentation of these events should be kept in an *Issue Log* that will be monitored like other control items. Someone on the project team should be charged with monitoring the issue log and following up on overdue responses.

To show how an issue can convert to a risk event, let's use the broken copy machine. Normally, the copy machine can be fixed by a local service copy in one day. What happens if the copy machine cannot be fixed and will require a long lead new machine? This situation starts to take on the face of an unknown/unknown risk event at some severity point and is handled through a more formal resolution process. Personal preference is to keep the role of risk management separate from minor daily events. On the surface, managing minor issues through the risk management process seems like excessive administrative work. Regardless, the management goal is to ensure these ad hoc items don't get lost and are handled on time.

# CONTROL ACCOUNT MANAGEMENT

The definition of a Control Account (CA) was discussed earlier in Chapter 8. The essential idea for this is to better support the monitoring and control of that work cluster target. A second aspect of the idea is to assign a formal manager, called a Control Account Manager (CAM), for each CA. This delegation provides an improved level of responsibility for tracking and managing project status. The key management strategy for this role is to place responsibility on the individual best prepared to manage that collection of work. The second aspect of this is to delegate and obtain more buy-in to improve work performance. One element of this structure is to recognize that the management of small WPs is not always practical, whereas it is more effective to manage some larger aggregation. This supports the idea of budget data collection remaining either at the WP level or be moved to the CA as logic dictates.

The formal use of CA and CAMs is not a well-used idea in practice, but it should be seriously considered. One of the behavioral characteristics that this approach deals with is the "your plan versus my plan" syndrome. The better goal is to think of it as "our plan." If the CAM buys into his part of the overall project, he will be more motivated to achieve the plan. Also, this formal delegation makes the execution responsibility more focused and hopefully improves the team partnership concept.

# RESERVES

Discussions related to reserve accounts have been mentioned in various other parts of the text. A reserve may be justified to cover variations in schedule, scope, risk, budget, and resources. One way to think about reserve usage is their role in handling a level of uncertainty for various process groups. In each case, the goal is to protect the planned status value for that variable. The recommended view of a reserve is to think of it as an estimating error and try not to mix that with a scope change or risk events. For example, a schedule reserve buffer says that the schedule may vary by the defined reserve amount from undefined causes. This is the level of variation that is projected but is not specifically identified. The same logic would be used for all other defined reserves. Once established, a management role is required for each reserve pool to evaluate the adequacy of each reserve as the project progresses.

## Schedule Reserves

In the case of an approved schedule, time buffers can be added to the plan to cover a level of forecast variation.

## Budget Reserves

Similar to the approach taken for schedule variance, a budget reserve should be created and status monitored throughout the life cycle.

## Risk Reserves

Chapter 9 has provided insights into the need for a contingency reserve related to risk events. Deciding on the amount to place in this reserve pool is considered an art form more than a mathematical exercise.

## Resource Reserves

Assuming that an evaluation of resource requirements has been made during the planning cycle, logic suggests that even with a good estimate, there will be a need for a different resource pattern for various reasons as the project progresses. Given the multiple threads of the resource picture, it is little wonder that this area is challenging to synchronize with the plan needs as the project unfolds.

## Scope Reserves

This chapter described the role of the ICC and the need for a scope reserve pool to handle approved changes in scope. Like the risk reserve, the size of this reserve is based on historical data or expert judgment.

# CONCLUSION

Successful control of any project is complex because of the wide variety of variations and related root cause factors. A mature status monitoring and control methodology is a primary key to success. The methods discussed here represent a reasonable summary of the core concepts used in the project management arena. Beyond the raw mechanics related to this activity, it is important to recognize that the collection of planned versus actual status metrics neither hinders nor enhances project status understanding. These comparisons passively describe what did occur. It is up to the management process to delve deeper into this data to understand the causal elements and then take corrective action. Once control variations have been identified through the KPI measurement approach, it will be the role of the project manager to decide which combination of methods and degree of data granularity will work best in a particular project. There is simply too much variety in the project environment to merely list a group of KPIs that will appropriately highlight project variability areas. However, this chapter has provided typical areas for this activity.

The concept of management control flexibility is an essential ingredient for effective control and one that many organizations do not always follow. In some situations, a heavily standardized KPI reporting requirement creates extra work on the project team and does little to improve the project tracking needs. For average-skill project managers and less technical projects, this may be a suitable strategy, but an overly structured approach can result in a less effective outcome for complex projects being managed by a highly skilled manager. It is important not to inhibit the project manager's decision-making flexibility to the point where there is no degree of freedom remaining in collecting status data for internal needs.

We will close this chapter with a final philosophical point for project control. If there is a variance in any aspect of the project, ask yourself why. Try not to leave the occurrence of a variance measure unexplained. Document the reason for the variance and record that in a lessons learned repository available for your project and others. The mantra of a learning organization is to omit negative events and repeat positive ones.

# REFERENCES

BrightGuage. Md, MSPs 70+Metrics and Formulas Available at: https://www.brightgauge.com/blog/quick-guide-to-11-types-of-kpis (accessed September 1 17, 2021).

Carstens, D.S. and Richardson, G. 2020. *Project Management Tools and Techniques*, 2nd ed. CRC Press, Boca Raton, FL.

Parth, F., and Guma, J. 2003. How project metrics can keep you from flying blind. Available at: http://www.projectauditors.com/Papers/Whiteprs/ProjectMetrics .pdf (accessed April 18, 2021).

Richardson, G.L., and Jackson, B.M. 2018. *Project Management Theory and Practice*, 3rd ed. CRC Press, New York.

# Project Team Management

# 13

## INTRODUCTION

This chapter builds upon project communication discussed in Chapter 10, but now focuses on optimizing project team management. As defined here, "The project team consists of individuals with assigned roles and responsibilities who work collectively to achieve a shared project goal" (PMI, 2017, p. 309). People do not awaken one day and decide to be ineffective and disorganized project managers or project team members. Generally, people have intentions to do a good job in their professional positions (Shetach, 2010). Not only is the project manager responsible for effective communication with all project stakeholders, as discussed in Chapter 10, but should also focus attention on retaining and improving team satisfaction and motivation throughout the project life cycle. Therefore, there are specific skills and competencies that the project manager should possess to manage their project team successfully. Different tools exist to help a project manager and project team be organized to maximize the overall team performance in terms of accuracy and efficiency.

## ORGANIZING PROJECT TEAMS

Project teams need to clearly understand their role within the team and the overall organization. Every team member's "output" is "input" for another team member's assigned task. Understanding this concept will help each team member identify how they fit into the project's overall scheme and how important each person's role is to the team. This model helps team members understand the importance of keeping up with their deadlines so that their output properly links into follow on tasks. If any team member is late with their output, it could delay the entire project because the next task can't begin without the predecessor team member's output. Eventually, all the defined output

DOI: 10.1201/9781003218982-13

generated leads to the successful completion of project deliverables. There are many situations where groups can effectively use an input and output relationship model to complete project tasks.

A relationship-type organizational chart should be developed for the project teams, so all stakeholders are aware of which individuals are responsible for different tasks. Figure 13.1 displays a sample organizational chart. Note the use of names under each box to help with the defined chain of command. This type of schematic provides an efficient method to show general responsibility and assists in problem resolution. This is just a sample of organizational documentation, and other similar formats can be used (see Chapter 5.)

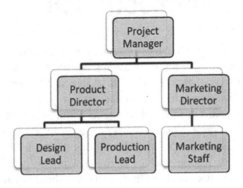

**FIGURE 13.1**    Project team organizational chart.

# PROJECT TEAM PERFORMANCE TOOLS

This section describes the different organization and communication tools not covered in Chapter 10. The focus is oriented towards performance management. Project managers need "skills to identify, build, maintain, motivate, lead, and inspire project teams to achieve high team performance and to meet the project's objectives" (PMI, 2017, p. 337). To successfully do this, project managers must be good communicators using open and effective communication. They also need to help team members develop trust with each other, which can be accomplished through team-building opportunities. The ability to resolve conflict is another strength a project manager needs because there will not be one or two such events but likely daily episodes to handle between various stakeholders. Table 13.1 is a tool that can be used to work through such events, with the project manager serving as the facilitator. Having a positively oriented project manager that involves appropriate team members is the proper method, and this form of decision-making is vital to the team's success. Another trait that a project manager should have is the ability to listen actively. This consists of "acknowledging, clarifying and confirming, understanding, and removing barriers that adversely affect comprehension." (PMI, 2017, p. 386). It is also important to have cultural awareness and ensure that all team members respect different cultures, ethnicities, and diversity. The organizational role of meeting management is also important. This consists of distributing agendas before meetings and communication of upcoming meetings.

**TABLE 13.1**   Conflict Resolution Template

| CONFLICT RESOLUTION WORKSHEET | |
| --- | --- |
| Identify the Conflict: | Identify the Stakeholders Involved: |
| Describe Options for Resolution: | Describe Consequence for Each Option: |
| Prioritize Each Option: | Select an Option: |
| Create an Implementation Plan for the Option Selected: | Assign Dates for Implementation Plan Tasks: |
| Signature of Stakeholder 1: | Comments by Stakeholder 1: |
| Signature of Stakeholder 2: | Comments by Stakeholder 2: |
| Signature of Project Manager: | Comments by Project Manager: |

There are four distinct phases of group development: forming, storming, norming, and performing (Tuckman, 1965). In the forming stage, teams form, and there is little known about the project goals. Team members may not know each other. The initial phase of team building is called *forming*. At this early stage, brainstorming can be used to develop an understanding of project objectives. Table 13.2 is a tool to assist in brainstorming project goals and tasks. Another communication format to use at this point is to request that each team member list on a sticky note one perceived task from the defined work specification. There will be several iterations of this as broad themes emerge. As this effort continues, tasks, subtasks, and eventually more specific Work Packages will be defined. Some would call this the "buy-in" phase for the team. The second stage of group development is called *storming*. During this stage, there will likely be many clashes as the team sorts out how they will work together, but handled properly, the members will create a trust relationship which will provide a valuable foundation for future phases. The third stage of group development is the *norming* stage. At this point, the role relationships start to become accepted and defined. Disagreements are worked out, and cooperation among team members emerges. The last stage is called *performing* because the team is now in work mode. If done properly, the work requirements are well understood, and the role of management is more of support and less of an authority relationship. Failure to accomplish this team growth process can lead to less desirable outcomes. Using traditional authority and power concepts is a suspect strategy for team success, especially when the project does not formally own resources. Table 13.3 displays a template for a team performance appraisal form.

**TABLE 13.2**  Project Brainstorming Template

| *BRAINSTORMING TEMPLATE* |
| --- |
| Brainstorm tasks needed to be conducted to meet the project goals. |
| Brainstorm input for the quality management plan. |
| What are the challenges for this project? |
| What are the project risks? |
| What are ways to mitigate the risks? |
| Who are additional stakeholders not yet identified? |
| What are questions not yet answered? |

It is important to share with the team when and why performance evaluations take place. Communication in this format is a major key to success, so it is essential to find methods that serve as timely performance input. Formal performance discussions are only one form of this, but an important one. A project manager's technique is motivating, and providing corrective inputs are important skills.

# SUMMARY

This chapter focused on the relationship dynamics of the project team. Project teams need to clearly understand their role in perspective to how each team member fits within the entire team. In the process of organizing project teams, a simple input and output relationship model helps team members complete their tasks and assists other team members with their tasks. A project organizational chart is also discussed as a tool to provide team members with an understanding of how each member fits into the overall project. In the project team performance tools section, the team development model was discussed (Tuckman, 1965). Other team development tools were provided, such as a brainstorming template tool and team member appraisal template.

**TABLE 13.3**   Performance Appraisal Template

PROJECT TEAM PERFORMANCE APPRAISAL

| TEAM MEMBER'S NAME: | | | PERFORMANCE REVIEW DATE: | |
| POSITION: | | | REVIEWER'S NAME: | |
| PERFORMANCE CATEGORIES/ | EXCEEDS EXPECTATIONS | MEETS EXPECTATIONS | BELOW EXPECTATIONS | COMMENTS |
|---|---|---|---|---|
| Job Knowledge | | | | |
| Productivity | | | | |
| Quality of Work | | | | |
| Attitude | | | | |
| Adaptability | | | | |
| Interpersonal Relations | | | | |
| Decision-Making | | | | |
| Overall Comments: | | | | |
| Team Member's Strengths: | | | | |
| Team Member's Weaknesses: | | | | |
| Recommendations for Development and Improvement: | | | | |
| Team Member's Comments: | | | | |
| Team Member's Signature: | | | | |
| Project Manager's Comments: | | | | |
| Project Manager's Signature: | | | | |

# REFERENCES

PMI (Project Management Institute). 2017. *A Guide to the Project Management Body of Knowledge*, 6th ed. Newtown Square, PA: PMI.

Shetach, A. 2010. Obstacles to successful management of projects and decision and tips for coping with them. *Team Performance Management* 16(7–8):329–342.

Tuckman, B. W. 1965. Developmental sequence in small groups. *Psychological Bulletin* 63(6):384–399. doi:10.1037/h0022100. PMID 14314073.

# Closing the Project

# 14

## INTRODUCTION

This chapter describes the need and processing steps to undertake in closing a project. Special emphasis is given to contractual closing, lessons learned to value, and the need for a team celebration. Leaving a project in an unclosed state is considered very poor management practice and can have a significant negative impact.

The PMBOK® Guide defines the Close Project process "as finalizing all activities for the project phase and contract" (PMI, 2017, p. 634). This activity closes out that portion of the project scope and associated activities applicable to each phase for multiphase projects. The closing process should be formal and documented. This defined set of activities includes actions related to project administrative and contractual issues. Closing the project also includes documenting lessons learned from the project so that future project teams can learn from the collective lessons from the project team.

## TARGET AREAS TO REVIEW

There are at least eight areas to review in the project as part of the closing process:

1. Vendor management
2. Equipment issues
3. Project approval process
4. Budget approval process
5. Communication issues (lateral and vertical)
6. Testing
7. Technical support
8. Training

DOI: 10.1201/9781003218982-14

In each of these areas, the concern is to document the current status and what needs to be done to complete the area. The following scenario examples will help justify this as a required activity:

1. Suppose you loaned equipment to a vendor to help with the contract. The vendor now thinks you gave it to them. This is the time to resolve that issue.
2. You have been under the impression that the contract was finished when all the tasks were completed, and the customer accepted the product. Senior executive made a verbal commitment to provide one year of training to the new users (yes, that happens). Now you have an unplanned requirement.
3. Like the item above, the contract specifies customer warranty support for the product for one year. Your team is the only group capable, but this was not included in the approved plan.

## PROJECT TERMINATION CHECKLIST

Horine (2005) outlines the following topic areas to be covered in the project checklist (pp. 282–284):

1. *Obtain Client Acceptance.* The client formally verifies and accepts the project deliverable, and this event is formally documented.
2. *Transition Deliverables to Owner.* The team formally hands off the project deliverables to the new owner. This includes possession of the item and the ability to support the item long-term.
3. *Close Out Contract Obligations.* The project team will coordinate with procurement personnel to document the status of all contractual relationships.
4. *Update the Organization's Central Information Repository.* This activity involves documenting project records and deliverables as a formal archive for the organization.
5. *Document Final Project Financials:* This includes a budget status summary and variance analysis.
6. *Close Various Accounts and Charge Codes.* This activity involves the process of closing team member accounts and codes related to financials, infrastructure, and security.

7. *Update Resource Schedules:* Work to ensure that team members have appropriate job opportunities following the closure.
8. *Conduct Performance Evaluations.* The PM must ensure that appropriate performance feedback is performed and documented for all team members.
9. *Update Team Resumes.* The team members should update their resumes to reflect the new project activity.
10. *Market Project Accomplishments:* Formally recognize team member accomplishments and overall project positive experiences.
11. *Review Project Performance with Clients.* This process evaluates whether the team achieved the desired goal.
12. *Celebrate.* From a team morale standpoint, it is important to find something to celebrate after a project concludes. Try to leave the project team feeling positive about their experience.
13. *Capture Lessons Learned:* Documenting the project team's experience-related activity enables future projects to profit from both good experiences and mistakes made.

The steps above have value not only for the termination phase but can also be of operational value as each life cycle stage is completed.

## CONTRACTUAL CLOSING EXAMPLE

It is one thing to have ambiguity in contractual terms that affect the project team; however, it is another to leave formal contractual terms unresolved. On the surface, this may not seem likely because the original terms are clear, but during the execution phase, items could have been sent back, failed in testing, lost, etc. One side thinks the item is clear, and the other disagrees. This scenario represents a subtlety that does not seem important, but it is likely the most dangerous area to omit in project closing. Recognize that after a project is completed, the team is often dispersed and, in some cases, leaves the organization. The basic goal of contract closing is to match actual events to documented contractual terms along with any addendums. This requires matching deliverables with fees, plus any other terms. Once the audit is complete, a formal meeting with the vendor should be held to review the status. That status should be summarized in writing and signed by both parties. Every effort should be made to leave the contract as clean as possible. Recognize that any conflict item is a potential lawsuit, so deal with this activity accordingly. It can be difficult to reconstruct all the machinations that have occurred throughout the life cycle. Once the project team has dispersed, the internal knowledge is also gone, so written documentation will be important.

# LESSONS LEARNED

One area of project management that now is viewed with increased interest is a formal lesson learned process. This statement is true for the current project, but maybe even of more value for future projects. Mature organizations have processes for internal evaluation of status and associated corrective action in areas that need improvement. The lessons learned process serves a major part of that role.

From an operational viewpoint, many of the project artifacts have sharing potential (i.e., Business Cas, Charter, WBS, estimates, plans, and templates of various kinds). Easy cross-access to this class of project documents can save considerable time for the following project.

# CLOSING CELEBRATION

This process is also important. In many cases, the project has been difficult and may even have been terminated early. The team needs to feel like they accomplished something regardless of the circumstance. Later, when team members get back together to share memories, you will hear a sentence starting with "Do you remember when?" View the project as a learning experience and one that is not so bad in hindsight. If all has gone well, this is easy but try to find something productive to share and congratulate the team members. The team closing celebration can be a simple pizza lunch, or an expensive couples evening out with full regalia. This get-together will generate a positive reminder that the project had some good aspect to it. Of course, a monetary performance reward bonus would be even better and should be considered. The management goal is to make the closing a positive event. If the project manager has done his or her job, the team will be able to look back and see that their professional expertise was improved.

# REFERENCES

Horine, G. M. 2005. *Absolute Beginner's Guide to Project Management.* Boston: Que Publishing.

Project Management Institute (PMI). 2017. *A Guide to the Project Management Body of Knowledge (PMBOK® Guide),* 6th ed. Newtown Square, PA: Project Management Institute.

# Agile Development

# 15

## The Iterative Approach

---

## INTRODUCTION

---

This section aims to describe the basic elements of a major growth trend in project management, that being a class of development called "agile." Several dialects of this methodology are now evolving, with the most popular one called Scrum, aptly named after a Rugby huddle technique. Proponents of this technique do not look favorably at the traditional waterfall method. It is important to understand why all project managers should understand and evaluate various management techniques to properly apply appropriate trade-offs between various methods. No one method is optimum for all situations.

Agile represents the family of iterative development, which was initially focused on software projects where unknown risks, high complexity, and rapidly changing project requirements were the common profile. The fundamental concept of this method fits into what one might call *lean thinking* because it shares similar concepts such as respecting people, focusing on faster-delivering value, and quickly adapting to change. The usage of this technique has now expanded outward beyond the initial starting place as certain design concepts have been proven to produce higher customer satisfaction in new profiles such as government, financial services, etc. Figure 15.1 shows a schematic that illustrates the iterative refinement process. Design requirements for the agile life cycle are dynamic instead of fixed, as in the waterfall design. Here, work plans change with each execution cycle as driven by customer interaction. In this format, agile involves repeating short development cycles with frequent increments of deliverables until the customer is satisfied. With each release, the customer provides feedback to the product owner along with feedback from the

DOI: 10.1201/9781003218982-15

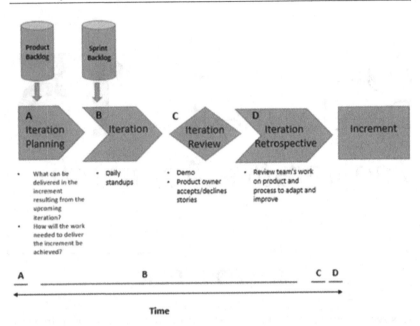

**FIGURE 15.1**   Iterative life cycle. *Source:* Richardson and Jackson (2019).

stakeholders. This feedback is added back into the process by being prioritized in the product backlog, which essentially is a wish list of requirements for the product. The high-priority items are selected from the backlog and placed on the sprint backlog (Development Pays, 2017; Straughan, 2019). This planned work effort becomes the sprint backlog as in the prioritized wish list of work to perform during the sprint and is the only item worked on during the sprint (Scrum Alliance, 2015; Development Pays, 2017).

Iterative development focuses on small increments (chunks) of work to gain customer feedback and change the future direction based on input from the previous activity to conquer uncertainty surrounding project requirements. This approach has been found to reduce waste and rework because the project team gets timely input from the customer. Note that the concept of scope change is built into the process rather than attached to it as in the waterfall approach. In this format, the team can respond quicker and more accurately than using a preplanned scope with formal written specifications, which often turns out to be inaccurate later. This approach to development is based on a mindset of collaboration with active customers who will use the ultimate deliverable(s). The short development work cycles focus on the team's ability to respond quickly and better understand requirements based on timely customer inputs. In this mode, the customer can review sample output and provide improved direction for the next work cycle. Similarly, this approach supports the team being better able to respond to changing project scope.

## Agile Manifesto

A team of software developers envisioned the agile methodological approach and described their management views in a document titled the 2001 Agile Manifesto (Hilbert, 2017). Included were four high-level *value* statements that set the tone for the team's interaction and 12 supporting *principles* that laid out more specific operational steps. First, recognize that agile is a blanket term that maps to various related dialect techniques that are similar in approach, method, practice, or framework based on the iterative concept (PMI, 2017).

The first manifesto value statement specifies that there should be a focus on "Individuals and interactions over processes and tools" (PMI, 2017, p. 8). Given this focus, the team will be more responsive to changes and better meet customers' needs. Focusing on people also instantly increased communication with customers instead of just at scheduled intervals throughout the project life cycle is considered more to be the traditional model.

The second value specifies "Working software over comprehensive documentation" (PMI, 2017, p. 8). The agile approach focuses on creating working deliverables in small segments using less formal specification documents such as *use cases*, *user stories*, etc. A *use case* describes situations regarding how the output will be used. The goal of these abbreviated specifications is to help get developers into a correct and realistic mindset. These somewhat prioritized mini snippets of requirements provide a very high-level scope definition. Each user story gives developers enough information to identify a general goal and a rough time estimate needed to implement it.

The third Manifesto value specifies "Customer collaboration over contract negotiation" (PMI, 2017, p. 8). This value describes how customers must be actively engaged during the entire project life cycle. In this mode, agile focuses on flexibility and customer engagement throughout product development so that customer needs are more readily and easily met. Customers are an active part of even daily meetings and interactions. When questions arise, the team can dynamically incorporate the needs into the upcoming mini design cycle instead of later requiring costly rework.

The fourth Manifesto value states that the project team must be flexible and quick in "Responding to change over following a plan" (PMI, 2017, p. 8). The traditional life cycle process creates detailed plans focused on outlining what a deliverable would do, how much time and budget would be required, risk assessment, and other process guidance intended to produce a robust roadmap to execute the complete effort. In the waterfall model mode, changes to the plan are not encouraged but rather controlled and handled as approved. Changes are viewed as disruptive to the original plan. The agile model iterative approach is the opposite, where changes are encouraged, and work is loosely defined before each work segment based on customer input. The design focus of agile is to optimize customer satisfaction with the final product while admittedly shorting some of the forward-looking visibility built into the traditional view.

Within the four-core value are 12 more specific *Manifesto principles* that further drive the process (PMI, 2017, p. 9):

1. The highest priority is to satisfy the customer through early and continuous delivery of valuable software.
2. Welcome changing requirements, even late in development. Agile processes harness change for the customer's competitive advantage.
3. Deliver working products frequently from a couple of weeks to a couple of months, with a preference for the shorter timescale.
4. Business representatives and developers must work together daily throughout the project.
5. Build projects around motivated individuals. Give them the environment and support they need and trust them to get the job done.
6. The most efficient and effective method of conveying information to and within a development team is face-to-face conversation.
7. A visible working product is the primary measure of progress.
8. Agile processes promote sustainable development. The sponsors, developers, and users should be able to maintain a constant pace indefinitely.
9. Continuous attention to technical excellence and good design enhances agility.
10. Maximizing the amount of work *not* done is essential (avoid waste).
11. The best architectures, requirements, and designs emerge from self-organizing teams.
12. At regular intervals, the team reflects on how to become more effective, and then tunes and adjusts its behavior accordingly (learning organization).

# ORGANIZATION STRUCTURE

The agile development environment does not have a predefined organizational structure with typical lines and boxes. Instead, it is multiple circles, with each containing a team placed around the leadership core. Servant leaders focus on the management of relationships within the team and the entire organization. This involves communication and coordination with all stakeholders, including mentorship, listening, and coaching roles; the collaborative environment approach allows teams to learn and become more capable. This process supports improved problem-solving and cross-communication skills. Team members must learn how to move quickly to define the why behind the work so that the proper next step can be more quickly established. The process focus is results-driven in terms of value delivered to the customer.

An agile team consists of individuals serving different roles that strongly depend on each other to deliver value to the customer (PMI, 2017). There will need to be cross-functional team members that possess the skills needed to produce the product. These skills align with the iterative life cycle so that those on the team can carry out all the life cycle components. A product owner is responsible for working with the team daily to drive the work direction of the team regarding which product functionality to work next. This role requires continuous involvement with all stakeholders to provide transparency into the activities of a team and in directing the team to focus on delivering value for the customer. It is understood that there will be incremental changes that the team will need to make to the product requirements as the features desired by the customer evolve. A team facilitator serves as the *servant leader* for the team. This role has different names depending on the methodology dialect, but some common names are *agile coach*, *Scrum master*, etc. It is important to supply an appropriate physical workspace for the agile team. This requires a mixture of collaborative workspace for the entire team and areas where the team members can work individually in an uninterrupted manner. It is also important to have proper support tools such as video conferencing and shared access to project documentation.

Similar to the traditional approach, an agile project must have a defined management approval process to get started. The minimum starting trigger is formal management approval for the effort and a Project Charter to provide the high-level project vision, purpose, release criteria, and the intended workflow. The servant leader role facilitates the necessary steps to obtain the project Charter.

# AGILE METHODS AND PRACTICES

The CollabNet survey (2020) offers an extensive view of agile practices in the marketplace, and it will be used here to quantify the goal focus related to this methodology. The *benefits* of adopting agile were (respondents made multiple selections) as follows:

70% to manage changing priorities
65% project visibility
65% business/IT alignment
60% delivery speed/time to market
59% team morale
58% increase team productivity
51% project risk reduction
50% project predictability
46% software quality
44% engineering discipline
41% managing distributed teams
35% software maintainability
26% cost reduction

On the counter side, the *challenges* identified by the CollabNet survey (2020) for adopting agile were (respondents made multiple selections) as follows:

48% general organization resistance to change
46% not enough leadership participation
45% inconsistent processes and practices across teams
44% organizational culture at odds with agile values
43% inadequate management support and sponsorship
41% lack of skills/experience with agile methods
39% insufficient training and education
36% lack of business/customer/product owner availability
30% pervasiveness of traditional development methods
29% fragmented tooling and product-related data/measurements
22% minimal collaboration and knowledge sharing
16% regulatory compliance or government issue

# SUMMARY

This chapter provided a brief background of the agile methodology concepts, including a brief history of its organization, reasons for adopting it, benefits, and challenges. Carstens and Richardson (2020) offer more details on related software support tools, and Richardson and Jackson (2019) describe time theoretical details regarding agile and other adaptive life cycles.

# REFERENCES

Carstens, D. S., and Richardson, G. 2020. *Project Management Tools and Techniques*, 2nd ed. Boca Raton, FL: CRC Press.

CollabNet. 2020. 14th annual state of agile report. Available at: https://stateofagile.com/#ufh-i-615706098-14th-annual-state-of-agile-report/7027494 (accessed June 26, 2021).

Development Pays. 2017. Scrum vs. Kanban – What's the difference? Available at: https://www.youtube.com/watch?v=rIaz-11Kf8w (accessed July 2, 2021).

Hilbert, M. 2017. The real origins of the Agile Manifesto. Available at: https://www.red-gate.com/blog/database-devops/real-origins-agile-manifesto (accessed June 25, 2021).

Project Management Institute (PMI). 2017. *Agile Practice Guide*. Newtown Square, PA: Project Management Institute.

Richardson, G. L., and Jackson, B. M. 2019. *Project Management Theory and Practice*, 3rd ed. New York: CRC Press.

Scrum Alliance. 2015. What is scrum? Available at: https://www.youtube.com/watch?v=TRcReyRYIMg (accessed July 2, 2021).

Straughan, G. 2019. Scrum vs Kanban cheat sheet. Available at: https://www.developmentthatpays.com/files/DevelopmentThatPays-ScrumVsKanban-CheatSheet-1_6.pdf (accessed July 2, 2021).

# Which Product Structure Is Right?

# 16

## INTRODUCTION

This chapter describes key input/output variables that guide the definition of a proper project life cycle structure. Each project goal is somewhat different, and these differences should be taken into account in designing the appropriate life cycle. The agile methodology has shown the project world various techniques that have demonstrated merit in improving project outcomes, yet not all projects fit this contemporary model. This chapter will outline the key techniques that need to be employed in all life cycles regardless of the underlying management approach, whether waterfall or iterative. One cannot simply grab a methodology off the shelf and universally apply it. It is not simply one model versus the other!

A previous chapter described how project teams get extremely ego-focused on their venture (i.e., their baby). The same thing can be said about project structure bias. Agile followers strongly believe that this approach is superior to the traditional waterfall structure that requires pre-definition and documentation of requirements and resources before management approval and execution. The industry should give credit to the agile school of project management to show how their modified design structure techniques improve customer satisfaction and cycle times. On the other hand, it is time to recognize that neither life cycle model is magic, and success is more related to understanding how to execute the appropriate requirement processes.

## COMMON PRINCIPLES

Unfortunately, a cross-comparison of waterfall versus iterative (agile) takes on an almost religious perspective. Once one approach has been adopted, all

DOI: 10.1201/9781003218982-16

other methods are viewed as inferior. The theory described here hypothesizes that both formats can be correct (or wrong) based on the project goal. Before starting this review, we need to understand that the agile model contains visible differences in scope definition, defined status metrics, and the associated control processes. Some of these techniques can also be utilized to improve the waterfall results. It is also important to recognize that basic agile does not fit all project types. Let's offer one seemingly clear non-agile example. Does it seem feasible to iteratively develop the construction of a bridge or a building? Although even here, the answer is not crystal clear. Within this project type there seems to be potential task characteristics that might fit best as an agile subprocess. Certainly, of a mixed hybrid structure view makes sense to consider. It would not seem too uncommon to envision many project types with this same potential. If so, how can one envision the correct design to contain both iterative and waterfall characteristics? In other words, what combination of structure and processes would be considered? On the surface, it seems appropriate to embed some design-oriented agile sprints and still represent the overall project using tools such as a WBS and WPs.

Beyond this high-level potential hybrid view, some general agile principles would seem appropriate to embed into all life cycle designs. The following list is a sample subset of general agile principles to consider:

- Actively involve stakeholders inside the project team for help in interpreting requirements, evaluating results, and feedback.
- Keep defined workgroups as small as possible and manage them like sprints. These could be called Control Accounts with the same meaning.
- Keep the team informed regarding the status using daily stand-up meetings.
- Communicate status and lessons learned.
- Avoid multitasking through the sprint packages.
- Avoid task procrastination and padded work estimates.
- Establish WBS chains (sprints) of tasks and execute them as quickly as possible.

Note that many of these are also well-known project management concepts, but they are often not standard practice in the traditional project culture.

The following statement may sound like agile heresy, but recognize that *"most of the agile principles are known techniques that good project managers should already be following."* There have been many sections of this text that have described these desired traits in the waterfall context. Iteration and prototyping do not have to be required to incorporate these. It is important to admit that some project types do not lend themselves well to iteration and prototyping, but the use of other agile principles would likely have the same positive impact on a hybrid model utilizing these principles. These are good management techniques and processes relevant to all model structures and goals. The key is to understand how to utilize each of these in a specific design. Here is one visible daily observation to help see this point. When you look at current road construction and see miles of orange cones, stacked-up traffic, and see no workers in sight, have you ever considered that there might be a better way to deal with the current defined requirements goal? Road construction fits the traditional model very well, but some agile "speed-up" would be appreciated by the motorist. Would smaller work units (sprints) produced quicker be a possible alternative here? Can you envision agile laid on top of the current model? Remember, there is no value to the new road until it is put into use.

A typical question remains, why not convert all projects to some dialect of agile? Our answer here is to suggest that we quit branding the project structure as a fixed model and start utilizing these common principles.

Managers who believe the answer to this question is found in such a cookbook are likely to have minimal success compared to those who can design a custom approach that optimizes the project goals and utilizes appropriate management principles. To accomplish this goal, some additions would be needed for both views. What modifications would be needed in an agile design as a customization example if the approval process requires an estimate of budget and schedule before approving the project? That structure would seem to require some level of requirements definition beyond basic agile. This could be resolved by adding various waterfall-like processes. The same type of additional modification logic goes for deliverable and resource specifications. Previous discussion has indicated that it is very hard to control project status on the control front without having some baseline metric defined, along with at least a milestone schedule and budget plan. These are all waterfall concepts regardless of the visible tool to be used. Yet another potential organizational requirement for project approval is some form of risk assessment, and this is also a handy waterfall process. Agile does not prescribe a risk assessment process, but it may well be a needed activity before execution for some project types. This additional activity could uncover risks in pursuing the effort, or at least indicate alternative approaches, even mitigation activity. One might argue that organizations should not have these pre-execution requirements, but they exist and are part of the common project culture that is often necessary to deal with. For the logic outlined here, it seems reasonable to suggest that the two best-known project structure approaches each has merit. The appropriate parts of each should be utilized to capture the best characteristics of both.

As stated earlier, any discussion of methodology cross-comparison seems to resemble more of a religious belief than one based on fact. Recognize that neither side of this debate is dumb, and neither is completely correct. The agile movement has demonstrated that changing accepted management and the technical process takes a long time to accomplish, but when these do occur, the flock often moves too much toward the newer side. Agile has taught the industry a great deal about some excellent techniques to improve project outcomes in a relatively short period. The goal here is not to take sides on these approaches but to show that proper project management has many commonalities regardless of the underlying design philosophy. These approaches should not be that different in actual operation.

# DESIGNING A CUSTOM STRUCTURE

An old vaudeville skit goes something like this—The patient says, "Doctor, doctor, my arms hurts." The doctor responds, "when does it hurt?" The patient says, "when I do this." The sage doctor then says, "Don't do that." One can learn a lot from vaudeville. As a project manager, you must do whatever is necessary to keep the project's arm from hurting. Utilizing good management and operational practices like the ones outlined in both agile and traditional life cycles is fundamental to that goal, but remember that nine conceptual management knobs have to be coordinated and tweaked to drive project results (i.e., product functionality, scope, time, budget, risk, resources. quality, communication, and stakeholders). Agile focuses primarily on a successful product and short completion schedule, but these are only two of the management knobs, albeit very enticing ones, to say the least.

The answer to the question of "what is the best product structure" could fill many reams of counterargument text. We won't take up this non-resolvable question, but choose instead to look at few organizational culture items that would help improve outcomes, maybe even more than defining a single fixed methodology approach. There is clear evidence that incorporating the following five items would do wonders for project productivity and outcome results regardless of the underlying methodology:

1. Recognize that there are many projects under consideration in an organization. Proper management requires a portfolio-type approval process, and this requires a formal data-oriented forecast to rank the options. Doing the wrong project well is still the wrong project.
2. Project customers are often tentative about being actively involved with the project, thinking that it is either too technical or they don't see that as their role. This should be a required aspect of the effort.

3. Resources are often not managed well in terms of timely availability or skill. Agile tends to resolve this by having a dedicated and task-focused workforce. Be sensitive to the fact that schedules automatically slip when resources are not available on schedule. A formal resource management process is needed to support this need.

4. One of the greatest evils in the traditional task work model is the padding of time estimates to avoid overruns. This leads to a culture of procrastination and multitasking that exacerbates the problem. Note what agile has done to minimize both of these factors. This same central focus on task execution culture is needed on all projects.

5. Budgeting and scope change concepts in the traditional model are confusing and error-prone. The traditional budget process generally does not deal well with scope change, and most organizations do not handle risk or other reserves well. There is too much buried in the task structure to know what is going on in this maze. Agile keeps its work structure open and visible. Recognize that budget management in an agile project is not a major factor. Some say that agile projects are finished when the money runs out. Think about the role of cost control in your particular product structure.

There is no clear answer to defining a custom project life cycle model, but we see some key goal parameters here to evaluate. The following sample factors are representative of items that guide the selection of the most appropriate management structure:

1. Does the organization require a forward plan outlining desired goals such as product, cost, schedule, deliverables, and risk?

2. Can project work tasks be reasonably defined before execution? If this cannot be defined, will the project be approved to start?

3. There are various project target goals such as product, schedule, budget, risk, quality, etc. The project plan design and subsequent management actions must focus on the correct one of these.

# THE FUTURE

There will likely be yet another modified answer to the appropriate life cycle question 50 years from now. It will also likely be driven by some new technology beyond our thought process today. Projects may be executed by automated robots; who knows. If you are serious enough to research what this future model might look like, take a look at Goldratt's Cortical Chain theory and ponder the relevance of that approach (Goldratt, 1997). Critical Chain theory has flavors of both agile and traditional theory, but it is packaged in a unique way that many organizations are not yet ready to handle, even though the base theory has been proven to work. However, the Critical Chain processes are very disruptive to the traditional management control approach, even more so than agile has been. Amid this ongoing search for the holy grail, there is a lot known about managing successful projects. The agile implementation has been educational in that it was conceived internally as a grassroots effort, but also note that this has taken 20 years to reach the current maturity and recognition state and is still evolving.

Our hope in this final text chapter is that you, the reader, have absorbed sufficient theory from the various process discussions to sort out the key issues required to manage this complex topic. The art of project management is much more complex than most understand, and it is not amenable to a singular cookbook approach. You have now been exposed to the major factors that correlate with project success/failure and the associated management strategies that support successful outcomes.

Good luck in your future attempts to produce successful project outcomes. We hope this shortened version has given you some worthwhile background and additional food for thought.

# REFERENCE

Goldratt, E. M. 1997. *Critical Chain*. Great Barrington: North River Press.